HOW AI THINKS

www.penguin.co.uk

HOW AI THINKS

How we built it,
how it can help us,
and how we can control it

NIGEL TOON

torva

TRANSWORLD PUBLISHERS
Penguin Random House, One Embassy Gardens,
8 Viaduct Gardens, London SW11 7BW
www.penguin.co.uk

Transworld is part of the Penguin Random House group of companies
whose addresses can be found at global.penguinrandomhouse.com

Penguin
Random House
UK

First published in Great Britain in 2024 by Torva
an imprint of Transworld Publishers

A CIP catalogue record for this book
is available from the British Library.

ISBNs 9781911709466 (cased)
9781911709473 (tpb)

Typeset in 11.5/16 pt Baskerville by Falcon Oast Graphic Art Ltd.
Printed and bound by Clays Ltd, Elcograf S.p.A.

The authorized representative in the EEA is Penguin Random House Ireland,
Morrison Chambers, 32 Nassau Street, Dublin D02 YH68.

Penguin Random House is committed to a sustainable
future for our business, our readers and our planet. This book is
made from Forest Stewardship Council® certified paper.

MIX
Paper | Supporting
responsible forestry
FSC® C018179

To Sally

CONTENTS

INTRODUCTION

'What we want is a machine that can learn from experience.'[1]

MATHEMATICIAN AND PIONEERING
COMPUTER SCIENTIST ALAN TURING, 1947

We are at the start of a revolution that is already beginning to shake our world: an artificial intelligence revolution. AI is the most powerful tool that humans have yet created, and we are just beginning to learn how potent it could become. It has the power to increase our own incredible human intelligence, helping us to solve complex problems that were previously impossible. But AI also brings with it many new challenges – profound changes lie ahead. Over the course of my life, during a career spent working at the forefront of technology, I have been directly involved in developing the underlying technology that makes AI work, and I have an understanding of how this story will play out. I want to help you learn how AI thinks.

It all started when I was around eight years old and my father brought home part of an old computer, thinking that I might find it interesting. As a talented electronics engineer who had worked on developing some of the early computers during the 1950s, he was able to show me how the buttons on this old switchboard were used to program these early machines. I was fascinated to learn that computers can't directly understand letters or words. Deep inside their electronic circuits they communicate using numbers, but the numbers that they use are strange. They don't use the decimal system we are all familiar with; instead they use a peculiar type called 'binary'. I started to realize that computers are very different from humans. How computers work, and whether it might ever become possible to build one that could be called intelligent, has fascinated me ever since. We tend to think of computers as complex calculators doing exactly what they are told, step by step in a program, but now these powerful machines are learning from information.

As you sit down to read this book you are being bombarded by a huge amount of information coming at you from every direction. You will be wondering whether the chair you picked is comfortable and deciding if the cup of tea you just made has a bit too much milk in it. You are perhaps being distracted by the view out of your window or by the background noise that's trying to grab your attention. Your brain is probably still buzzing from the conversation that you just finished, and you may also be trying to remember that event coming up next week. Most likely there will also be that nagging feeling that you have forgotten to do something and that you shouldn't just be sitting here reading a book. Your brain will be trying to filter all these different pieces of information to work out whether any of it is useful, whether any pieces are connected, if something needs to be remembered, and what to ignore – while at the same time trying to focus your attention on reading this text.

Over the course of a day your body burns around 2,000 kilocalories, averaging about 100 watts of power every hour. This is about the same amount as is needed to light an old-fashioned filament light bulb or ten new LED versions. When riding an exercise bike, you might peak at 200 or even 250 watts, unless of course you are a professional cyclist like three-time Tour de France winner Chris Froome, who regularly exceeds 400 watts sustained over a 30-minute period.[2] When asleep, your power consumption will drop to around 25 watts and right now, while sitting reading this book, you are at around 50 watts, with roughly half being consumed by your brain. This means that you will need to eat around 500 kilocalories – roughly equal to two chocolate bars or five apples – to make it to the last page. You are about to turn food into knowledge.

As I communicate with you through the pages of this book, I hope to share some useful information. As I will try to show, the communication of information can increase our knowledge, which in turn provides us with more intelligence. Communication is critical for building intelligence, but the information that can help us may come along at any time. Often, we will need to store it away so that we can use it later. We may also need to generalize this information so that we can apply it to different situations. It turns out that memory is just a time-shifted form of communication that can provide useful information later, when we need it most. You heard something that sounded interesting six months ago, stored it away, and finally the moment arrives when it's useful to tell someone else. Communication, information, and the knowledge that results, all help us build intelligence. Humans first used this intelligence to find food and now we use it to improve our lives in a host of other ways.

AI, too, relies on the communication of information, something that was made much easier by the opening-up of the internet. AI requires some of the most advanced and specialized computers, which have been provided for us through incredible advances in semiconductor chips.* It also needs humans, to develop new methods that allow these powerful computers to learn from this digital information. The result is an intelligent machine that can recognize objects, create images that never existed before and even hold a conversation with you.

Our ever-growing store of human knowledge has given us the intelligence to build the advanced tools and machines that we use every day to make our lives easier. Humans are good at building

---

* Conductors, such as metals, pass an electric current, while insulators, such as plastics, block the current. Semiconductors are a special type of material that can either pass or block an electric current, depending upon their state. Silicon is an example of a semiconducting material and is actually one of the most common materials on Earth, which you will find as sand on every seaside beach.

machines – the engines that we build increase our human strength, allowing us to work harder, go faster and even fly. But now, with these new AI methods, we are building machines that can learn from information to capture more knowledge that will in turn help us increase our own level of intelligence.

Artificial intelligence is the most powerful tool that we have ever created. The ability to amplify human intelligence will let us solve problems that we previously found impossible. But it is only in the last few years that truly powerful AI has finally started to emerge. AI is a completely new approach to computing. Over the coming years, these new intelligent machines will transform our lives – indeed, the transformation has already begun.

As AI has emerged, this amazing technology has also become the subject of many dinner-party conversations and public debates. Some supposed experts are stepping forward to express strong opinions about the challenges and disruptions that this new intelligent force may bring. But artificial intelligence is complex and so, like other powerful tools that have come before, our lack of understanding can easily lead us to think that AI might present a threat to humanity. When confronted by a new technology, we are prone to throw up our hands and say we just don't understand. We need to learn more, not just about the risks but about the benefits that this important new technology can deliver, because it will end up changing our lives and, perhaps more importantly, the lives of the next generation.

Following my early encounter with a computer, as a teenager I went on to build my own machine using an early microprocessor and became fascinated by how it worked. Forty years later I started building a business that developed a new type of microprocessor that accelerates artificial intelligence computation. Over the course of my career, I have been fortunate enough to work at the cutting edge of semiconductors, computing, communications and AI. I have had access to the boardrooms of the largest semiconductor companies, toured the leading silicon-chip fabrication plants on the island of Taiwan and in Korea, worked with companies on optical communications systems that send the internet around the world, and on smartphone technology that puts the internet in our pocket. I have visited super-computer labs in the UK and USA, and together with my company co-founder and friend Simon Knowles, I have met many of the people who are driving AI forward. This includes such leading figures as Demis Hassabis, the founder of the AI research company DeepMind; leading academics such as Geoffrey Hinton and Yann LeCun, both Turing Award winners for their breakthrough work in AI; Stanford University computer science professor and former head of the Google Brain project Andrew Ng; plus Professor Jürgen Schmidhuber, who has been one of the leading innovators in AI over the last three decades. During these conversations and in many others, innovators have provided me with an espresso shot of their subject-matter expert-ise, but to try to build an even broader understanding, my search for knowledge has led me to look in detail at neuroscience and biological systems, both of which inspire AI systems.

Lots of people ask me about artificial intelligence, including friends, family, doctors, lawyers, business leaders, civil servants, politicians, our UK prime minister and even members of the British royal family. Many have said that they don't understand AI, that it scares them. The idea that a machine can make decisions, and can reason in a way that we thought was unique to humans, makes them worry about what AI may become capable of in the future. Our lack of understanding leads us to fear the machine.

I have written this book to try to address these concerns. I want to arm you with knowledge so that you can benefit from the enormous opportunities that AI will deliver. To do this I will try to show how AI thinks and describe the technologies that have made it possible. We will explore how AI can help us, as well as considering the controls that we will need to put in place. I will also try to explain how machine intelligence differs from our own very special human intelligence.

This fear of the machine is understandable. During the first test-screenings of the animated movie *Shrek*, many children became very anxious about the early renderings of the film's heroine, Princess Fiona – she was too lifelike, and children cried when she came on screen. This phenomenon in animation and robotics was first identified by Masahiro Mori, a robotics professor from Japan, in a paradox that he called the 'uncanny valley'.[3] Mori showed that as robots or animated characters are made to look a bit like humans, people will initially respond positively and think they are cute. However, as the robot or animation starts to look more lifelike, this previously positive feeling can very quickly turn to unease, with the character becoming creepy. We are confronted not with an ogre that shows humanity, but a human mutated into an ogre. This effect is also played on by the directors of horror movies, where very human characteristics are given to non-human characters such as dolls – for example, in the films *Dead Silence*, *Annabelle*, or *The Boy*. Many dystopian movie plots also play on this human worry and feature a sentient AI that is trying to take charge. This uneasy connection between humans and human-like machines is perhaps what lies at the root of our fear of AI. If we could learn more, then hopefully we will come to understand that, like any machine, AI is just a tool that is following a method imprinted by humans, has limitations in what it can do and is under our control.

Another more subtle point relates to data and how we learn. Data on its own doesn't help us. Without context, data is meaningless. A picture on our smartphone is just a set of dots until we can learn that these dots are showing a picture of a cat. Most of us would probably already know what a cat looks like; we have this knowledge hidden away in the neurons of our brain – but how did this knowledge get there? At some point in our very early life, perhaps when we were first introduced to one, we were told, 'This is a cat.' This description provided us with the context that we needed to identify this unfamiliar creature. Context turns *data* into *information*. After seeing a few more cats we were able to generalize these different pieces of information and make associations that helped us build a deeper understanding of cats. Through a process of learning from information, we built *knowledge*. With this knowledge we could then start to show off our new *intelligence* by pointing out a cat every time we saw one.

As the famous Cole Porter song goes, 'Birds do it, bees do it, even educated machines do it . . .' Well, not quite, and computing machines certainly don't fall in love, but as I will share, animals, plants and even single microbial cells follow this same organic process of using data plus context to capture information. This information then lets them build knowledge, and from this knowledge they gain intelligence.

And so now, rather than telling a computer what to do, step by step, in a program, humans have come up with reliable methods that let computers learn from information too. These AI methods use information to build an AI knowledge model from which the machine can start to make decisions that exhibit so-called 'artificial intelligence'.

To help you understand how AI thinks, I have tried to write for anyone who has a natural curiosity. I want to make this complex subject accessible for people who do not necessarily have experience in maths, science or technology. This is my attempt at sharing some of my knowledge so that others can gain a deeper understanding of the significant changes that artificial intelligence will bring to our lives. I hope that time spent trying to understand the most transformative technology of our age will open incredible new possibilities.

To do this we first need some basic understanding of AI. Along the way, I will also share some insights into human intelligence so that we can compare and contrast. It's actually very hard to find a clear description of intelligence that everyone can agree on. The *Oxford English Dictionary* defines it as: 'the faculty of understanding'. However, like most definitions, this is narrowly bound and open to interpretation. As a result, when a machine does exceed human intelligence, such as the IBM Deep Blue machine beating world champion Garry Kasparov at chess in 1997, we often either dismiss it as a 'trick' or recategorize this 'machine ability' as something that falls outside our definition of intelligence. We are also often surprised and concerned when AI produces strange answers, as BBC News recently reported. Their article[4] highlighted that an experimental AI-powered chatbot built by the online giant Meta was providing unexpected replies when asked about the company's founder, Mark Zuckerberg. 'His company exploits people for money and he doesn't care. It needs to stop!' the chatbot said. So, if we are going to consider whether AI is intelligent, we first need a better understanding of intelligence and a more robust definition. As the pioneering computer scientist Alan Turing said, '. . . if a machine is expected to be infallible, it cannot also be intelligent'.[5]

To help build our understanding, I will share insights on the underlying technological developments that have allowed us to build AI, introducing some of my personal heroes, many of whom have helped to make our modern life possible. Gaining some insight into the amazing developments in semiconductors will help us see how AI technology has become possible. I will try to briefly outline how a computer can learn from information to deliver solutions that show a level of what we can clearly define as intelligence. But this background knowledge will also help us understand why artificial intelligence is not going to run out of control. We will look at how computers were originally developed, with software built on top, and how, over the last seventy-five years, they have evolved to a point where we are now starting to deliver on the promise – which the original computing pioneers such as Turing believed possible – of building intelligent machines.

In Part 2 I will start to tackle the question of how AI thinks, and to do this I will share some fascinating comparisons with the intelligence that has evolved in humans and other living species. We will dive into the difficult subject of consciousness to explore whether a machine might ever become sentient. We also need to understand whether AI could ever grow so powerful that it will start to reprogram itself and become progressively more and more intelligent, leading to a so-called singularity, an intelligence explosion that could end up making humans obsolete. Much has been written on such hypothetical Doomsday scenarios, but as we will see, there are good reasons to be optimistic about the future.

Artificial intelligence is already beginning to transform our lives. In healthcare, in finance, in transport, and in the workplace, AI is delivering incredible breakthroughs. In Part 3 we will look at some of the ways that artificial intelligence can help us solve problems that have been out of reach until now. Humans have been making and using tools for millions of years and our attitude towards them has often turned from fear to acceptance (followed by dependence). However, we must look in detail at the risks that this powerful new tool might pose and explore how we can mitigate these. We need to give thought to the socio-economic impacts, to bias and personal privacy, and to smart weapons – all of which AI may make worse. We will explore ways in which we can build responsible AI and what controls need to be put in place so that these new AI systems can be used safely.

Our purpose as intelligent biological life is to use energy to convert information into knowledge, which in turn lets us build intelligence. Intelligence allows us to survive and prosper as a species. I hope to show how artificially intelligent machines will help support us in building more intelligence, and as a result can have a very positive impact on our economies and on our own individual lives.

The famous natural historian and broadcaster David Attenborough has been quoted as saying: 'The fact is that no species has ever had such wholesale control over everything on Earth, living or dead, as we now have: that lays upon us, whether we like it or not – an awesome responsibility. In our hands now lies not only our own future but that of all other living creatures with whom we share the Earth.'

Just like the earliest flintstones used for cutting and hunting, like the oxen-and-metal ploughs we used for working the fields, or like steam engines that first powered machinery, AI provides us with a new tool that can make our lives better by deepening our understanding of our world. It is an extremely powerful tool, the most powerful we have ever built, and so we should all try to understand how it really works and how it thinks. But what I will also try to show is that artificial intelligence is a tool that *can* be kept firmly under human control. AI will help humans to become even smarter and allow us to live up to what David Attenborough calls our 'awesome responsibility'.

In this book I will also make use of another innovation that has helped Homo sapiens build sophisticated technology, inspire and control one another, and ultimately achieve global dominance over every other species on our planet: telling stories. The human brain responds well to a good story, especially when we can relate it to our own experiences and to our understanding of the world. A story helps us capture meaning from experience.

As I learnt from my first encounters as a child, computers don't tell stories. Instead, they communicate by sharing something called binary digits (also called 'bits'). But now, like humans, they are starting to recognize patterns, to learn from information, and even, in a manner of speaking, to think. Those who take the time to understand *how* artificial intelligence thinks will end up inheriting the Earth.

Written during 2022 and 2023 in Somerset and London, UK; Palo Alto, California, USA; and Beijing, China; plus a few other places around the world, and on the trains and planes that travel in between.

HOW AI BECAME POSSIBLE

I've come up with a set of rules that describe our reactions to technologies:

1. Anything that is in the world when you're born is normal and ordinary and is just a natural part of the way the world works.

2. Anything that's invented between when you're fifteen and thirty-five is new and exciting and revolutionary, and you can probably get a career in it.

3. Anything invented after you're thirty-five is against the natural order of things.

DOUGLAS ADAMS,
The Salmon of Doubt[1]

1

THE AI REVOLUTION HAS ALREADY STARTED

HELLO CHATGPT

You have probably already experienced first-hand the growing power of AI technology. The AI revolution actually started over ten years ago now, but engineers and researchers had been working on developing artificial intelligence systems for many decades before that. In just the last few years we have seen both an explosion in capability and in public interest, combined with a rising level of concern. A big wake-up call for many was the launch in November 2022 of ChatGPT, a powerful AI language application built by leading AI research company OpenAI. Fed with just a short text prompt, this incredible tool will generate lucid documents and answer complex questions. It comes close to meeting the requirements set out by Alan Turing in his 'Imitation Game'[1] test. Turing claimed that if a machine could ever match a human in a question-and-answer conversation then it should be considered intelligent. When I first started 'chatting' with the ChatGPT system back in late 2022, it was easy to see just how far AI had come.

Following the launch, interest in AI exploded. After just two months, over 100 million people had tried the system, making it the fastest-growing application ever. But even the latest version of ChatGPT is not perfect, and you can still tell that you are talking to a

> Q *Write a press release about the discovery of the moon being made of cheese, include quotes from a cheese industry expert about how this changes everything and from a local mouse who is now building a rocket to get to the cheese moon.*
>
> A In a groundbreaking discovery, scientists have confirmed that the moon is made entirely of cheese.
>
> According to the research team, the moon's composition was determined through extensive analysis of samples collected during the Apollo missions, as well as more recent lunar missions. The findings have sent shockwaves throughout the scientific community, with many experts hailing it as one of the most significant discoveries in recent history.
>
> "This is truly a game-changing discovery," said cheese industry expert Sarah Johnson. "The potential implications for the cheese industry are enormous, and we are already seeing a surge of interest in lunar cheese production."
>
> Local mouse and aspiring space explorer, Jerry, is equally excited about the discovery. "Ever since I was a little mouse, I've dreamed of reaching the moon," Jerry said. "But now that I know it's made of cheese, I'm determined to build a rocket and get there as soon as possible."
>
> The discovery of a cheese moon is sure to have far-reaching implications, both for the cheese industry and for space exploration. Stay tuned for further updates as this story develops.

machine. Importantly, it still makes mistakes. But this was the first AI system that could engage in a very plausible conversation. ChatGPT is good on technical subjects where the scope is limited, but for some conversations you need to be aware that it will make up facts and might give you the wrong answer. As users quickly found, it has a habit of acting a bit like an overconfident 'stochastic parrot',[2] by which I mean it will generate coherent-looking text but without any obvious understanding of the underlying subject – blithely stringing together words, but occasionally getting things wrong. Unfortunately, it never shares any doubts that it may have in the veracity of its answers, and instead it will present incorrect information in the form of an assured statement.

To its credit, OpenAI built in content-moderation rules from the start that are designed to stop ChatGPT from generating rude or violent content, and from supporting illegal activities, such as answering 'How do I make a bomb?' But some people started collaborating on forums to share prompts and new approaches that tried to 'jailbreak' ChatGPT and make it bypass these rules. OpenAI monitors these approaches and is working hard to block these 'gaslighting' attempts, but this is a powerful tool and great care is needed.

Controversy had surrounded these powerful chatbots since at least June 2022, when an engineer called Blake Lemoine, who was employed to test the competing Google 'large language model' project LaMDA (now released under the product name Bard), claimed that the system was acting in a 'sentient' manner. He said, 'If I didn't know exactly what it was, which is this computer program . . ., I'd think it was a seven-year-old, [or] eight-year-old kid that happens to know physics.'[3] His claims became public when he went to the press after being dismissed from Google.

It has already become clear that these systems will have an enormous commercial impact. Microsoft has invested billions into OpenAI to support their work on AI language models and has now added ChatGPT to its Bing search engine, a competitor to the popular Google search application. Following this announcement, their Bing app downloads jumped tenfold. This conversational AI system posed a massive challenge to Google and its virtual monopoly in search. Google immediately called a 'code red' with their engineering staff and rushed out their competing Bard product. Unfortunately, an error generated by the Bard system, shown in the first promotional video, caused a $100bn knock to their company value.

AI is turning search engines into intelligent-answer engines. But many in industry and in government are concerned that we are just going to replace one monopoly with another. Some are calling for an open-source development project that would make large language models available to industry and governments in Europe and other countries, to ensure that this type of technology doesn't just sit in the hands of a few dominant foreign companies. Instead, they want leading-edge technology that can be made available to all, on open terms, and which could support their digital sovereignty. Unless this is done soon, the technology will have moved on and they will never be able to catch up, becoming dependent on either US or Chinese tech.

Today's generative-language AI systems are so good they can even pass the legal bar examination.[4] The OpenAI GPT-4 system not only passed America's multiple-choice Multistate Bar Examination, but also its open-ended Multistate Essay Exam and Multistate Performance Test, scoring a combined 75 per cent across the whole Uniform Bar Examination – not only good enough to pass, but placing it in the top ninetieth percentile. But large language models are not yet perfect. They work well on these rote learning tests but still lack common sense. They continue to make incredible progress every year, but it's important to realize that we already live in a world that runs on artificial intelligence.

AI IS ALL AROUND . . .

Although I work with AI and with technology every day, I am still constantly surprised by new developments, and it's easy to feel overawed by what might become possible. You probably use lots of advanced computing technology in your daily life already. Some sits in plain sight, but much is hidden. Today, most complex products are powered by advanced microprocessors – not just in your smartphone, but in your coffee machine, inside your car (which actually contains hundreds of powerful microprocessors) and even in your purse or wallet. Hidden away inside thousands of remote data centres, most of which have been built within the last decade or two, you will find huge amounts of computing power that quietly makes the complex online technology, which we all now take for granted, just work.

In your home, you probably already own a powerful robot that saves you time and helps with your daily chores. Maybe this comes as a surprise, but a robot can be defined as an automated machine that performs tasks with little or no human intervention. Using this definition, your washing machine certainly qualifies as a robot. This seemingly pedestrian appliance contains some very advanced technology. Modern washing machines have a powerful microprocessor hidden inside that not only controls the front panel and interprets the switches and buttons that set the machine program, but also powers the wash cycle, working to reduce energy consumption and limit the amount of water consumed. The machine's motor is carefully controlled by a separate microprocessor programmed to make the drum rotation as smooth and energy-efficient as possible. Recently, a new washing machine was launched that advertised the fact that it has special sensors for monitoring the laundry's weight and level of soiling, then uses artificial-intelligence techniques to optimize the amount of water, detergent and rinsing time, making this new intelligent robot even more efficient.

Modern cars are also already full of artificial-intelligence technology. My car has a snazzy feature that recognizes speed-limit signs and tells me what speed I should be driving, warning me when I am going too fast. It senses when a car is overtaking me on the motorway and beeps at me if I attempt to pull out at the same time. It detects when I run near the edge of the traffic lane and flags if I get too close to the car in front. In an emergency it would actually help to vigorously apply the brakes to avoid an accident – luckily, a feature I have not yet had to test. These sophisticated driver-assistance features have all become possible due to recent advances in artificial intelligence.

I can tell that the intelligence in my car is artificial because very occasionally the image-recognition system will get the speed sign wrong and tell me that I am in an 80mph zone when it's actually 30mph – needless to say, improvements are still needed. I didn't sign up for the expensive self-parking feature. Instead, I get lots of confusing pictures and warning alarms that shout when I get too close to an obstruction. My car even told me once that 'parking is not possible', which of course sparked my competitive spirit. I did indeed manage to squeeze into the space, accompanied by a chorus of deafening parking alarms from the car. Unfortunately, my car doesn't have a way of learning from experience, and so it still complains every time I try to repeat my success.

In January 2023 Mercedes announced a 'Level 3' autonomous car (though initially only available in Nevada, USA). A Level 3 autonomous vehicle can perform all normal driving functions but still needs the driver to stay alert and be ready to take control when necessary – a bit like sitting next to an inexperienced driver. It won't be long, though, before we have fully autonomous 'Level 5' cars, which can operate without any assistance.

Your credit-card transactions are also now powered by AI. Hidden inside your card is a semiconductor chip that includes a special security microprocessor that connects to the payment machine either through the metal connector on the front or via a low-power wireless connection. Your card securely stores encrypted information, such as your credit card number and your card-holder information, and as you make payments your card provider builds up a detailed picture of your shopping habits. Artificial intelligence is being used to help confirm that it is actually you that is using your card. I travel a lot on business and have had situations where I will be using my card in multiple countries during the same working day. I have often received a call or a text from my card provider to confirm that it is indeed me making the payment or whether my card has been compromised. There have also been times when my card has in fact been misused, and so I am glad that AI has helped to detect these events.

When surfing the Web, you are triggering activity on lots of different servers in data centres all around the planet, leaving behind a trail with your digital footprint. For example, when you use a search engine you are firing up a number of different computers that will handle your enquiry. The search engine page that you land on will be powered by one machine, part of which will become dedicated to your enquiry. The search term that you enter will then be sent off to another set of machines that today use the most advanced artificial-intelligence 'natural language processing' (NLP) techniques, such as large language models, to help make sense of your query. The search results that you see will be different from any other person's and will be based on your previous search history, your location, the device that you are searching from and even your mouse movements on the screen. All this computation burns lots of energy, and each search, which takes about 100 seconds of compute time – or what computer experts like to call just 'compute' – consumes about 10 watts. That doesn't sound so much, but with over 10 billion searches being performed every day, that adds up to a lot of energy. AI is now being used to try and optimize power consumption in these data centres to make them more efficient.

At the same time, companies are competing to get their information shown in your search results, not just in the adverts that feature at the top of the page but also in the other information that's suggested. Companies put in lots of effort optimizing their Web content so that it ranks as high as possible in search results. Once you find something of interest you are taken off to yet another server, often in a completely different data centre, where you get to see the information that you selected.

The major search companies, big online retailers and social media companies are all making massive investments in artificial intelligence and are pushing the limits of what is possible. They are trying to get you better search results, show you products that more closely match your buying habits, or deliver content that you might find more compelling. But they are also using AI to encourage you to spend more time on their platform. Every time you use one of these online services you are engaging in a battle of wills with an AI machine. They are using AI to try to hold your attention, make their service more tuned to your personal preferences and keep you engaged.

Every month these AI systems are being improved. Research into the latest AI methods is often led or funded by these large online companies and it is this research work that has driven much of the recent progress in artificial intelligence.

The AI examples that I have described so far are still quite simple, even ChatGPT. The OpenAI GPT-3 language model, which ChatGPT was originally based on, has around 175 billion parameters – which are equivalent to the synaptic connections that structure information in your brain. GPT-4 has around 1.7 trillion, which is still small when compared with the 100 trillion or so that you have in your brain. The next generation of AI systems will drive fundamental changes in many of the products and services that we use every day, and the potential of this new breakthrough technology is enormous.

We stand on the brink of a historic change that is set to reshape our society. This creates an enormous opportunity for those who understand something about this powerful new technology. To start learning how artificial intelligence thinks, we need to explore the ways in which AI differs from traditional computing approaches and understand more about our own human intelligence.

2

INTELLIGENT MACHINES

This year, around 150 million new intelligent machines will be delivered. Each one has the potential to be much more intelligent than you. During your lifetime they will end up setting laws, and will advise and even lead governments. These new thinking machines will be running our businesses and will also start many new ones. They will beat our current sporting champions in every event. In your old age, you need to hope that these new intelligent machines will look after you, because in time they will control everything – they will end up in charge. We call these intelligent machines: children.

When a baby is born it already knows so much. It knows how to feed, how to see, how to hear; it will start to recognize objects and very quickly will know your face. Soon it will develop motor skills and start sitting up, then it will begin walking around, exploring its world. You will show it your cat and it will almost instantly be able to recognize any cat, even an animated version on a screen. Your children will also frustrate you, because despite your constant repetition, they will learn to say 'cat' before they say 'Mummy' or 'Daddy'. By dropping and bouncing a ball they will quickly grasp how gravity works, even if they don't understand any of the physics involved. It is amazing how much children know, how much they can learn, without even being taught, and how quickly they can become much more intelligent and more capable than you or me.

By comparison, computers are very stupid and don't have any of these skills. We need to tell them what to do at every step. When I was a teenager, computers were expensive, so with some help from my father, I built my own machine. It came as a kit, and I soldered the semiconductors and connected it all together, adding a keyboard, an old TV for a display, and a cassette recorder for program storage. You could also buy a chip that was pre-programmed with a basic operating system, but this cost extra so I decided to try to make my own. When the machine was turned on it just sat there. The keyboard didn't work, the display was blank – nothing happened.

To get the machine going, I had to build a little gadget that plugged into the empty 'read-only memory' socket, where the operating-system chip was supposed to go. I borrowed another computer and, by connecting this to the gadget I had built, I was able to get the machine to first show something on the display and then connect to the keyboard. Eventually I was able to write a very simple program that would scan the keyboard and write characters on the screen. It took a lot of trial and error to get the tape cassette working, but eventually I could save and load a small program. The computer needed to be told what to do, step by excruciating step.

We don't need to load an operating system, or add any other software, to make children function. Animals like humans are intelligent machines that already come preloaded with an incredibly capable operating system and lots of very sophisticated and intelligent software applications that all work seamlessly together. Much of our human software appears to be stored in our DNA and is passed down and developed through evolution.

Our DNA is built from a combination of information that comes from each of our parents, together with some random changes that are thrown in for good measure by the evolutionary process. These changes help us to evolve as a species and ensure that siblings and even twins are all different – this diversity gives species a better chance of survival. Scientists have estimated that humans have around 37.2 trillion cells,[1] each containing a total of 3 billion base pairs of DNA.

These two linked strands of DNA wind around each other and look a bit like a twisted ladder – a shape that is known as a double helix. Each base pair can be encoded with two small pieces of information, so a strand of human DNA can theoretically store 6 billion pieces of information, converted by 'gene expression' into physical action. Gene expression determines what function a cell should perform – for example, be part of your skin, a neuron in your brain or a cell in your immune system. We are not yet sure what much of it is used for,[2] but DNA appears to work like software that's inherited and passed down through our genetic line.

With my computer built, I was ready to write a program. I decided to try to make my own version of a simple 2D game called Space Invaders, which was originally developed by a Japanese engineer, Tomohiro Nishikado.[3] In this game, alien spaceships work their way

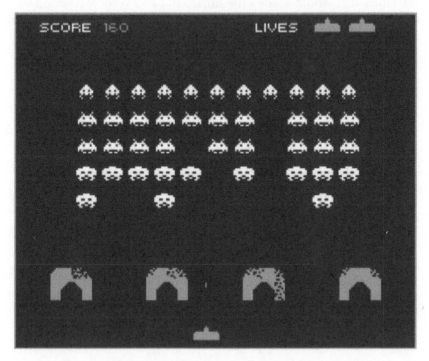

To program Space Invaders, you can divide the game into subroutines

down the screen trying to invade the planet at the bottom. The player controls a 'phaser' to shoot the alien vessels, but the aliens also shoot back. There are four blocks at the bottom of the screen that you can hide behind, but as the aliens shoot, pieces fall off until eventually there is no protection left. The purpose of the game is to destroy all the aliens before they destroy you. If you manage to clear the screen of aliens, the game starts again but this time the alien ships move faster, making it harder to win at the next level. It's a classic shoot-'em-up arcade game.

It wasn't hard to figure out how the game worked and how to construct the program. The key was to break the program down into different pieces, or subroutines, that could be called upon from a controlling program. I wrote subroutines to draw the alien spaceships on the screen, the phaser traces, and different versions of blocks – with and without damage. The spaceship subroutine had to include two variables that would specify the horizontal and vertical position for where the subroutine would draw the spaceship on the screen. By changing these variables, the spaceship could be moved around the screen, but the same subroutine could be reused to draw other spaceships in different places (computer-game developers call this type of subroutine a 'sprite'). Once these subroutines had been created, it was straightforward to write the control program that scanned the keyboard so that I could control the movement and firing of the phaser. The control program would also move the ships across and down the screen, and worked out what had been hit. Creating all the different subroutines that made up the program was laborious, but once the basics were in place, it was quite easy to change the speed of alien ships as they moved across the screen and to change the way the phasers worked to make the game better to play. I was able to improve the game by just updating the control program, rather than having to rewrite all the software code every time.

Humans are also very programmable. Put simply, we are much better at learning than a computer is. For example, when you hit a tennis ball, something very similar to my software program is happening. Muscles are controlled via motor neurons in the spinal cord that fire off commands to extend or contract your limbs. Your brain has a hierarchy of muscle movement subroutines that sends commands to these motor neurons to execute different sequences of muscle movements. You are not working out all the different movements, or directly controlling every muscle action and then changing all the code for the next tennis shot. Instead, you start with a high-level command that then triggers a set of pre-motor subroutines in the brain that in turn plan the correct movement sequences. These then control another hierarchy of motor-control subroutine that converts these movement sequences into a set of coordinates for what you need your limbs to do. This more precise information is in turn sent to the part of the brain that controls the motor neurons that trigger the correct muscle movements. It sounds complicated, but I have in fact massively simplified all of this because the actual mechanisms are even more complex. When you take time to train at tennis, you are learning how to improve all these subroutines. A tennis coach would call this 'improving your muscle memory'. With enough training, hitting a tennis ball becomes a subconscious act. Hitting a topspin forehand across court will take just a single thought that triggers the appropriate set of subroutines, which in turn control all the muscle movements necessary to make the shot happen. You actually need quite a lot of subconscious intelligence to perform this simple act of hitting a tennis ball.

THE INTELLIGENCE OF TENNIS PLAYERS

Imagine that you are standing on a tennis court and your opponent has just served a tennis ball at around 50mph towards you. It dips over the net, bounces off the court surface and your job is to hit the ball back using your racket, calculating the strength and direction of your return strike so that the ball will go back over the net and land inside the court on your opponent's side.

Any tennis player will roughly estimate the ball trajectory, the effect of the bounce and the impact that any spin might have, and will try to get themselves in the best position from which to strike a return. A good player will be able to place their shot down the edge of the court so that it lands just inside the sideline to win the point. A combination of your eyes, your vision system and your brain must somehow work this all out in about one second, which is the time it takes for the ball to cross the net and reach you as you try to set up for the return.

In a professional tennis match the serving speed is often well over 100mph and, as a result, the time to react and move into position for the return shot reduces to less than half a second. Scientists from the École Polytechnique Fédérale de Lausanne (EPFL) in Switzerland,[4] in the hometown of tennis star Roger Federer, have shown that amateur players actually have similar reaction times to the professionals, so the professionals must be doing something differently. Their research showed that professional players are often much better than the rest of us at certain time-related, perceptual skills. This includes recognizing the speed and direction of a moving object. These skills were obviously highly prized by early Homo sapiens to help us hunt. Now, instead of recognizing people for their great hunting skills, we appreciate and reward people that excel in these capabilities as our sporting champions. We may not be used to thinking of this as intelligence, but it is an impressive feat of calculation.

It might be possible to build a robot that could play tennis well, but it would end up being very complex and therefore very expensive. It turns out that there are some things that appear to be much easier for humans and some things that machines are better at. Contrary to our traditional assumptions, high-level reasoning actually requires very little computation, but low-level sensorimotor skills require enormous computational resources. While we find it quite simple to learn to hit a ball or ride a bicycle, and find it harder to solve a cryptic crossword or win at chess, the amount of compute required is actually much higher for the seemingly simple tasks.

This observation, known as Moravec's Paradox,[5] was identified in the 1980s by artificial intelligence and robotics researcher Hans Moravec, who concluded, 'It is comparatively easy to make computers exhibit adult level performance on intelligence tests or [in] playing checkers, and difficult or impossible to give them the skills of a one-year-old when it comes to perception and mobility.'

I am not very good at tennis, but I do enjoy trying to play. Just occasionally I will run to the ball with my racket outstretched, connect with the ball in just the right way and send the ball down the line for a great return. When this happens, I am always a little shocked because there was no obvious conscious thought involved. I was just reacting to the flight of the ball – everything else came from my muscle memory. I happened to have just the right set of subroutines loaded and for once managed to get them all to work correctly. It's a great feeling and gives you a taste of what it must be like to be a well-trained tennis champion.

The author Malcolm Gladwell, who focusses on social science insights, famously wrote that if you practise at something for 10,000 hours then you can become an expert.[6] Unfortunately, I don't think this is correct. I believe that our neurons are wired a certain way at birth. Some of us will have more aptitude for a particular activity than others. We can still train and improve, but some of us will be born with more skill than others. The electrical pulses all pass through our bodies at approximately the same speed, but when they arrive they will trigger subroutines that are a bit more precise, a bit better controlled. For some people, getting a great result will come much more naturally. And, because things come more naturally, this will encourage naturally talented people to train more, and they will then get even better.

All humans are amazing and the difference between people's ability is probably quite small. Our tennis example gives us a clue. The professional player can return a serve that is sent at 100mph, whereas an amateur is only comfortable returning one that comes at 50mph. The difference between good and great is maybe about double. We might guess that someone who is not very sporty is a bit worse than our amateur player, but perhaps with enough practice that gap could be closed. The same is probably true between a genius and someone who is just clever – maybe the difference is only double, certainly not tenfold. Homo sapiens are actually all very similar. The difference in DNA from one person to the next is only about 1 per cent. Even though we might look different, have different-coloured skin, come from different parts of the world; though we might find that some of us are a bit better at sport and some are more skilled at maths or language, we are all very nearly exactly the same. We are all programmed with the same basic information, we can all learn, and we can all improve.

LEARNING TO LEARN

But can computers learn? For me, this question was clearly answered in 2015 when the journal *Nature* published an article written by engineers at DeepMind, an artificial-intelligence research company based in London, titled 'Human-level control through deep reinforcement learning'.[7] This article was based on work that had originally been published at the NeurIPS AI conference in 2013 and showed how DeepMind had managed to create a method for training a computer so that it could learn how to win at computer games. Not only did the method allow the computer to learn how to play the game without any human help, but it also allowed the system to reach superhuman levels of skill, learning tricks that only an expert player would know. The AI method used was called a 'Deep Q-Network', and the only inputs it received were pixels from the screen and the game score. It was not told how to play, other than the basic objective of the game.

Its only control was an emulation of the joystick movements and the firing button. Across a set of forty-nine different video-arcade games (that included Nishikado's original Space Invaders) for the Atari 2600 game console, the AI system was able to match or surpass the scores of human gaming experts. The artificial-intelligence method included a pattern-matching approach that could recognize the elements on the screen, as well as a way to interpret whether it was winning or losing. Equipped with this reinforcement learning technique, the computer was clearly learning 'how to learn from experience'.

Both humans and animals use a form of reinforcement learning to learn from experience. We try to understand the environment around us using different sensory inputs and then learn how to control this environment through trial and error. Learning through experience involves trying to generalize different approaches by analysing actions that work well and by understanding actions that have worked less well. The key is not to just *memorize* a specific situation but to *generalize*, so that this learning can be applied in situations that look similar. You are very unlikely to confront this exact situation ever again and so this *exact memory* is not very useful. However, if you can learn to *generalize* and apply the rule to situations that are similar, then you can make much faster progress.

Again, our tennis analogy works well here. A (right-handed) player might learn that a backhand topspin works well when the ball is arriving on their left-hand side at a height that allows the return shot to clear the net even with topspin applied. A player might learn to *generalize* that this is a good type of shot to try in this situation, but if instead the ball is dropping, it might be better to try a backhand slice that lifts the return up over the net. These first two rules make the player predictable, so we need a third rule: mixing these shots up from time to time will wrong-foot their opponent, allowing our player to win the point. Although each tennis shot is different, the player is able to *generalize* and builds up a portfolio of return shots that will work well in certain situations. Executing the shots will be a result of triggering one of the muscle-memory-trained subroutines that they have practised (*Slice!*). Learning which shots to use will come from generalizing a good game-playing strategy. By learning through experience about the opponent's weaknesses, they will also learn to try shots that their opponent might struggle to return.

If we had enough time and we had complete control over our neurons and nervous system, then perhaps we could all become an expert at anything. Here a computer has an advantage. When we create a machine learning method – a way for the machine to learn from information, rather than being told what to do, step by step, in a program – we have control over the whole machine. The machine never gets bored and can repeat the exercises again and again, making tiny improvements with every few goes. Even if things start to go wrong, the method can go back and try a different path. In the case of DeepMind's Deep Q-Network, it took about 50 million training steps to reach expert level on Space Invaders. Playing a 2D computer game is perhaps a bit easier than learning tennis, but even so, let's assume that after just 1 million half-hour training sessions even I could reach a professional standard at tennis. That would take me 500,000 hours, or about sixty years, of continuous training, which would be pretty intense. However, since humanity's ancestors have been on Earth for about 2 million years, it's no surprise that we have a head start, and some humans have evolved sensorimotor control skills at a level that allows them to excel at sport in far less time.

The challenge we are faced with is that the way in which our brains learn is not well understood. We know that neurons create new connections, and this appears to be related to learning. So-called neuroplasticity allows our neural connections and our nervous system to adjust its activity in response to internal or external events. We appear to reorganize the structure, connections and functions of our neurons as we learn.

We also know that humans can create new cells; in fact, we create billions every day[8] and continue to do this throughout our lives, although this appears to slow down as we age. If you cut your finger, your skin can heal itself by producing healthy new cells. In the same way, we also create new neurons through a process that is called neurogenesis. This process allows our brain to heal and to adapt. More and more research is being done to understand how we can keep our brains healthy,[9] and as our population ages this is becoming increasingly important.

A major factor in this process is rest and, in particular, sleep. Various research has shown that when we sleep our brains resolve information that it has learnt during the day, working out what was useful and what was not. Our brain is perhaps considering what knowledge to *generalize* so that it can become more broadly applied. A lack of sleep will lead to confusion and reduced cognitive function. Teenagers, who are going through a period of rapid growth and development, appear to need even more sleep as their brains begin to integrate valuable information about the adult world. Tired, grumpy teenagers appear to be a perfectly natural phenomenon.

Your brain is amazing, with more computing power than any computer that has yet been built. It consumes roughly the same energy as your laptop and is small enough for you to carry around with you wherever you go. You need to look after it.

*

With the computers that today's semiconductor integrated circuits have allowed us to build, we are already able to create methods that allow AI machines to learn from information. In some areas, machine learning systems are developing superhuman capability. Tasking computers with winning at games, which are closed worlds with clear rules, is just a way for us to understand the best methods for allowing a machine to learn from information. If we could apply these methods to other areas, we might be able to create tools that could help us solve problems that humans currently find impossible.

We are starting to discover that humans and computers work in very different ways. We now need to take a step back and understand how computers became so powerful, and how it became possible to start building the impressive AI systems that we have today.

3

THE BIRTH OF AI

Even though research has shown that only 12 per cent of Americans identify as 'cat people' and 42 per cent say they prefer dogs, on the Web people clearly like cats best. Cats are described as the 'unofficial mascot of the Internet'.[1] This observation was confirmed by an early AI research project in June 2012, run by a team from Stanford University and Google, and led by Stanford computer science professor Andrew Ng. They published a paper that described a neural network they had built, running on 16,000 computers all connected together in a data centre.[2] This system was set to work looking at images captured from YouTube. The system ran continuously for three days. Without being given any labels or clues, and without knowing what it was looking at, this very early 'deep' learning system discovered that there are way more images of things we describe as 'cats' on YouTube than anything else.

Modern AI systems use something we call 'artificial neurons'. The artificial neurons that AI uses are a much-simplified approximation of our human-brain neurons, as no one knows exactly how human neurons work. The artificial deep neural networks that are being used today are made up of millions of these artificial neurons. A typical system will include a set of input neurons that collect information (for example, the information from a digital picture), and a set of output neurons that provide a result (for example, recognizing a cat). In between are a network of artificial neurons arranged in layers that filter and mix the information as it flows through. These hidden layers are not programmed in the traditional way but are instead trained. For example, they are shown lots of pictures of cats until the network can start to reliably generalize and recognize any cat. The hidden layers of neurons learn to recognize a hierarchy of generalized features (called 'parameters'). These parameters are filtered from the input information and are then mixed and filtered again to build up even higher-level parameters. For instance, they may learn to recognize the outline of eyes and ears, which can then be combined to recognize a face. Ultimately, through more layers of artificial neurons, higher-level parameters learn to recognize complete objects (*This object is furry and has four legs*, for example). Eventually the deep neural network can learn to make a decision about whether there is a cat or a person in the picture. The answer that this approach produces is never 100 per cent accurate but is instead a 'probable' answer. But if well designed, these artificial neural networks can become extremely accurate – even more accurate than a human.

As an example of what has become possible, a group of researchers from the Icahn School of Medicine at Mount Sinai in New York and from the Keck School of Medicine at the University of Southern California have worked together to produce a deep learning AI system that can distinguish between low- and high-risk prostate cancer purely from scan images. This system helps doctors learn what intervention might be required. Doctors can now more quickly identify the treatment options and reduce the number of unnecessary surgical interventions. But AI technology developments have taken a long time to reach these breakthrough moments.

*

The term 'artificial intelligence' actually dates back to 1955, when John McCarthy, a computer science professor at Dartmouth College in America, proposed a summer workshop that would study a new approach to computing that he called artificial intelligence:

> We propose that a 2-month, 10-man study of artificial intelligence be carried out during the summer of 1956 at Dartmouth College in Hanover, New Hampshire. The study is to proceed on the basis of the conjecture that every aspect of learning or any other feature of intelligence can in principle be so precisely described that a machine can be made to simulate it. An attempt will be made to find how to make machines use language, form abstractions and concepts, solve kinds of problems now reserved for humans, and improve themselves. We think that a significant advance can be made in one or more of these problems if a carefully selected group of scientists work on it together for a summer.[3]

With hindsight, the goal of trying to solve the whole problem of AI in just two months was never going to be achieved. However, this summer conference did open up the whole field of AI research. Throughout the 1960s there was huge excitement about what could be achieved. Human-level intelligent machines seemed just around the corner. But these early research efforts also made the mistake of pursuing a research path that turned out to be wrong.

In thinking through a problem, we are usually trying to *infer* an answer from the relevant information, and from the retained knowledge, that we have available. We often post-rationalize why we made a specific decision. We tend to gloss over the fact that we didn't really have all the information or knowledge that we needed to make a strictly logical decision – hitting a tennis ball is a simple example of this. If we get the decision wrong, hopefully the outcome doesn't hurt us or anyone else, and we learn to do better next time. Making decisions and learning from them helps us build up experience, which in turn allows us to make better decisions next time.

The Greek philosopher Aristotle (384–322 BCE) described two ways in which we infer answers:

- Deduction, which is a process of reasoning from some statements to reach a logical conclusion; and
- Induction, where a set of observations are synthesized to reach a general conclusion.

In deduction the answer is certain, because it is based on some known facts and the decision can follow a clear, logical process.

For induction the answer is probable, because you are basing the decision on some information, but you don't necessarily have all the information that you need. You cannot make a completely logical decision.

If you are applying for a loan, you will hope that the decision process used will be based on deduction. It should use the information that you provided on your application form, and you hope that it does not rely on hidden information or bias. You would like to think that there is a clear logic that can explain how the decision was made.

However, when you need to make decisions as part of your everyday life, some hard facts will be missing. If you are renting a new apartment, there may be some very logical reasons for picking a particular property. But there will also be lots of unknowns, and you are unlikely to have all the information that you will need to make a purely logical deduction. So, you use the information that you do have – about the rental price, the knowledge that you have built up about the local area, information that you have learnt through experience of dealing with estate agents – and add in some assumptions and guesses. This all combines into a process of *induction* that will provide a *probable* answer. Most decisions that we make require some form of induction.

The early attempts at machine learning tried to use 'logical deduction' because this fitted much more closely with the existing software model of writing a logical step-by-step program. However, deduction will only work in a very narrow set of cases. You will need to have all the information to hand so that you can come up with a completely definite answer.

Some problems that appear to be solvable using deduction can actually end up being incomputable. This means that even though you have lots of information, however long you wait, a computer will just take too long to work through all the possible combinations to come up with a definite answer – the problem is just too complex. Deduction lets a computer control the anti-lock brakes on your car, for example, but cannot autonomously drive it across a big city like Tokyo.

To try to get around this limitation in deductive approaches, researchers in the 1980s attempted to develop more narrowly defined 'expert systems' that would limit the deduction to areas where lots of facts could be collected. These expert systems tried to capture knowledge from an expert person. As an example, researchers sat down with specialist consultant doctors and got them to share details about their area of medical expertise. They then tried to build an expert system that could be used by a general practitioner to get answers to questions that would help in a diagnosis. They hoped that this expert system could provide a distillation of the answers coming from these specialist consultants. The problem was that they underestimated human experts. The amount of information required, and the number of decision branches needed to cope with all the different questions that might be asked, grows very quickly. There are just too many possibilities, and the size of the problem quickly explodes so that finding an answer in this way takes an impossibly large amount of information and computing power.

Another example of this type of problem is board games. There appear to be only a finite set of possible moves, but in complex games like chess, and the strategy game Go, a very large number end up being possible. The mathematician and information theorist Claude

Shannon (who we will meet in Chapter 7) wrote a paper in 1950 called 'Programming a computer for playing chess',[4] in which he estimated that there are of the order 10^{43} possible positions in chess (that's a one with forty-three zeros after it). After just five moves by each player from the start of a game, there will be 69 trillion possible board positions for the chess pieces. So, to exhaustively search all moves, or even to try to look ahead by just five moves, would take far too much computing power. The deceptively simple game of Go is even more complicated. In fact there are far more possible moves in a game of Go than there are atoms in the whole universe. A modern AI system that is able to beat a world champion at chess or Go can't deduce what moves are possible: instead it learns how to develop a winning strategy.

This mathematical complexity problem was explained by Kurt Gödel in his 'incompleteness theorems', published in 1931.[5] These describe the limitations and incomputability of formal systems. Alan Turing's famous paper describing his 'universal machine', which later came to be called the 'Turing machine' and formed the basis for modern computers, also focusses on this issue. Turing showed that a machine applying a sequence of instructions to a set of information can perform any operation, but that some operations might take an infeasibly long amount of time. Gödel called this problem the '*Entscheidungsproblem*', or the 'decision problem'.[6]

Despite these challenges, some researchers continued to work on building AI systems, but instead of using *deduction* they switched to start working on building systems that could learn to make decisions by using *induction*. They started to focus on biologically inspired approaches and began developing artificial neural networks that would learn in a similar way to the human brain.

The very first 'deep artificial neural network' was developed by Kunihiko Fukushima from Japan in 1979, and the concept was then developed throughout the 1980s and 1990s, with teams in Canada and Switzerland being particularly active. In 2011 the first deep neural network, named DanNet after one of its creators, Dan Cireşan, beat a human on a simple image-recognition task.

Then, in September of 2012, a research team from Canada over-whelmingly won the annual ImageNet Large Scale Visual Recognition Challenge. ImageNet is a competition that was established to help develop computer-imaging systems and this was the first time that an AI system had won this competition. The winning AI model was called AlexNet, having been developed by Alex Krizhevsky in collaboration with Ilya Sutskever (who went on to found OpenAI) and their supervisor, computer science professor Geoffrey Hinton from the University of Toronto, together forming another renowned AI research team. These breakthroughs, dating from over a decade ago, kicked off the rapid progress in AI that we have seen in recent years.

How computers can help us was discussed by Steve Jobs in the early days of Apple Computer. In a 1980 video-recorded talk, he described a study he had read on the efficiency of locomotion for different species of animals.

> The condor [used] the least energy to move [a kilometre], and humans came in with a rather unimpressive showing, about a third of the way down the list. But fortunately somebody at *Scientific American* was insightful enough to test a man with a bicycle and a human with a bicycle won – twice as good as the condor, all the way off the [charts] . . . As a tool maker, man has the ability to build a tool that can amplify an inherent ability that he has . . . What a computer is to me is it's the most remarkable tool that we've ever come up with, and it's the equivalent of a bicycle for our minds.[7]

Artificial intelligence systems that use induction have finally started to deliver on the full potential of computers – to become a bicycle for our minds by amplifying human intelligence.

The early attempts at building AI were held back not only by the deductive learning methods that were used but also by the early computers. Early computers couldn't deliver enough performance to make AI work and the software we had was still rudimentary. There was also a lack of digital information for the machines to learn from. To learn how we have been able to build the advanced computers that now make AI possible, it will help if we spend some time looking at how modern computing technology first developed.

4

THE TECHNOLOGY
THAT BUILT AI – PART 1:
ELECTRONIC COMPUTERS
AND LEARNING TO SEE

On 5 June 1944, at a country house in the English countryside, General Eisenhower was in conference with his generals and staff, debating when to launch the D-Day landings. During this heated discussion a courier arrived from the Government Code and Cypher School at Bletchley Park and handed Eisenhower a top-secret piece of paper intended for his eyes only. On reading the message Eisenhower immediately announced to the group: 'We go tomorrow.'

The secret message that Eisenhower read had been intercepted and decoded from a heavily encrypted radio transmission sent directly from Hitler to Rommel, who was then commanding the troops responsible for the Nazi Atlantic Wall. Writing in 2006, Tommy Flowers, a pioneering electronics engineer at the time, recalled that Hitler's message had said that the invasion of Normandy was imminent, but that this would not be the real invasion. It was a feint to draw troops away from the channel ports, against which the real invasion would be launched later. Rommel was ordered not to move his troops. He was to await the real invasion, which could be expected five days after the Normandy landings.[1]

Eisenhower was not able to share the details of this message with the other senior people in the room. The fact that the Nazis' most secure 'Lorenz' encryption technology had been broken was top secret, known only to a very few. The insight that this decoded message provided gave Eisenhower a clear five days to land enough troops and equipment, and build a beachhead, so that he could then resist the inevitable counterattack.

It was just a few months earlier, on Saturday, 5 February 1944, that Tommy Flowers had written a short note in his diary:[2]

Colossus did its first job. Car broke down on way home.

This simple diary entry marks the birth of electronic computing, a technology that has gone on to power our technological society. Colossus was the machine that made the decoding of the German message possible. It was a breakthrough that would remain secret for nearly forty years.

During the 1930s, decades before transistors and semiconductor chips, Tommy Flowers pioneered research into the use of valves for electronic circuits at the telecoms research centre in Dollis Hill, London. Electronic valves could provide a replacement for the slow and unreliable electromechanical relays that were being used at the time. As Flowers observed, the use of electronic valves for fast circuit-switching 'was still in the research domain and was known to very few people in very few places in the world. By chance Dollis Hill was one of those places'[3]– and Flowers was one of the world's leading experts on the use of valves.

Flowers was called on by the code-breaking team at Bletchley Park to help Alan Turing with an improvement to his relay-based 'Bombe' decoding machine, which had been used to break the 'Enigma' codes that the Germans were using to encrypt their military messages. He was then asked to improve a new code-breaking machine that had been developed by Max Newman and his team to decrypt the much more challenging 'Lorenz' code used by the Nazi High Command. Flowers believed that he could build a much faster, all-electronic machine using valve technology but the team at Bletchley Park thought this was impossible. Valves are extremely unreliable and often fail, especially when the power is turned on and off. If you try to use lots of valves together, the chances are that one will fail and the whole system will then stop working. From his earlier research, Flowers had learnt that if you keep the valves powered up then they become much more reliable, and so he knew that his concept could work. He was able to convince Gordon Radley, who headed up the Dollis Hill research centre, of the importance of the project. Radley recognized Flowers' brilliance and the critical nature of this effort, and so he gave Flowers and his team his full support and they went to work in early 1943.

By December, a prototype system was running at Dollis Hill using a staggering 1,600 electronic valves. As one of Flowers' colleagues recalled, 'he didn't care about how many valves he used'.[4] The first system was shipped to Bletchley Park on 18 January 1944. The machine was massive, filling a large hall, and was dubbed 'Colossus' by the operators. This first Mk1 Colossus machine worked so well that Bletchley Park immediately requested more.

For the second machine, Tommy Flowers and his team had already planned several improvements. The Mk2 featured 2,400 valves, all working together as a single electronic computing system. This Colossus could be programmed via a special board on the front that had lots of plugs and connectors, allowing different operations to be configured. Information was input via a hole-punched paper tape. The system included electronic storage for intermediate information that was needed during processing, and it was able to process five separate streams all in parallel. This meant Colossus Mk2 was over a hundred times faster than the previous system. The speed-up allowed extremely complex encrypted codes to be decrypted within a few hours (or sometimes in minutes) rather than days, and provided timely intelligence for the Allied forces.

Flowers and his team had been in a race against time to get these improved systems built. He didn't know why the date was so important but Flowers had been told that they must be ready by 1 June 1944. The team worked six- and seven-day weeks from February to May to get the Mk2 machines developed and built on time.

On 1 June the Colossus Mk2 systems ran their first tasks and on 5 June the message that Eisenhower read was decoded and delivered. Subsequent messages helped to change the course of the war and undoubtedly saved many lives. Colossus was the world's first electronic programmable computer. No movie has yet been made of these dramatic events, but Tommy Flowers was the right person, in the right place, at just the right time.

After the war, Flowers continued his work on electronic switching systems, a technology that would ultimately transform the telecoms industry. During the 1950s, electronic computers started to become available as commercial products, and with the arrival of semiconductors they went on to deliver the incredible performance that modern computers give today. It was Tommy Flowers who built the first machine and he is a true pioneer of computing, whose work needs much wider recognition.

Computers began to perform calculations that now far outstripped human ability and they had begun to retain information. This would eventually allow them to learn how to see.

LEARNING TO SEE

Being able to take great pictures and videos of your friends in a dimly lit bar is one of the key design goals for a modern smartphone. Your device will perhaps feature two, or even three, camera sensors that all work together to capture the scene at different exposures. Your smartphone will then stitch these inputs together into a picture or video frame. The 'computational camera' system that sits behind the sensors in your smartphone is extremely complex, but it is still a long,

long way behind the amazing mammalian vision system that biological evolution has created for us.

A few quick tests will highlight this. Take your finger and hold it up in front of your eyes, and while continuing to look straight ahead, move your finger to the side. Most people with healthy vision should be able to still see their finger even when it is moved almost in line with their ears. If you struggle to see it, just wiggle it a little and it will jump into view. When you do turn your head from left to right, you can easily capture all the details from the scene around you over a full 360 degrees. Being bipeds also raises our vision system up above the bushes and undergrowth so that we get a much better view of the surrounding landscape. You can quickly scan the scene in front of you and then focus in on objects that are far off in the distance. You will notice even the smallest movements happening anywhere around you. Your eyes can also adjust from bright sunlight to very low light environments, and experiments have shown that a person with average vision can see a single candle flame that is well over 2 kilometres away[5] (assuming, of course, there are no obstructions).

Each of your eyes contains around 150 million photoreceptor neurons, with about 6 million 'cones' located at the centre that recognize colour, and around 144 million 'rods' seeing only in black and white but sensitive to the smallest movements. Your eyes are not static (like many insects' eyes are) but are instead constantly moving around to build up a complete picture from the scene, focussing in on the key subjects and quickly noticing any changes. A phenomenon called 'persistence of vision' means that your eyes only need to update the scene twenty times per second. However, your eyes can recognize a new object that emerges, anywhere across your whole field of vision, in just thirteen milliseconds.[6] You send all the details of this new object to your brain so that you immediately update your understanding of what is happening. You are able to react to this new information within a few hundred milliseconds.

'A picture is worth a thousand words', or so the adage goes, and humans are indeed much better at remembering images than they are at remembering words. We have a remarkable ability to identify pictures, and in tests people were able to remember 2,000 separate images that they had seen for only ten seconds each. Three days later, they could still remember these images with over 90 per cent accuracy.[7] Even when the viewing time was reduced to just one second each, people's recollection was not seriously affected. Over half of our brain is taken up with visual processing,[8] and this system is almost fully formed at birth, tuned by millions of years of evolution.

The visual cortex in mammalian brains has been studied[9] to try and understand how visual inputs from the photoreceptor neurons are filtered and pooled to build up higher-layer information from the scene. In a biological neural network, input neurons such as the photoreceptors in the retina each receive information about the world around us, such as light, shade, colour and movement. These input neurons are connected to the visual cortex in your brain, where other hidden layers of processing neurons work to understand and decode this visual-input information. Through hidden layers of connected neurons our brain builds up a picture of our environment and can then infer an understanding of the world around us, such as recognizing cats, dogs, people or the flight of a tennis ball. (You also have output neurons, which can activate muscles and trigger actions such as petting a cat or striking a tennis ball.)

This understanding of our visual system has provided the inspiration for artificial neural network systems that we call convolutional neural networks (CNNs). It was not until the 2010s that enough computing power became available, and the internet provided enough information, for deep neural networks to start to work effectively. Then, at last, the theoretical concepts for a deep learning vision system, inspired by how humans see, began to deliver accurate results.

The term 'convolution' is just a technical term for finding specific information and is sometimes referred to as a 'filtering operation'. In a CNN, layers of filtering are combined with layers of 'pooling', which, as the name suggests, is a process for combining and understanding this filtered information. To train these deep, artificial neural networks, we use very large sets of information that have been labelled so that the CNN can know what it is being asked to recognize – a bit like flash cards. Then, by passing this training information many times forwards and backwards through the deep learning CNN model that we have created, the system can be trained to reliably recognize different types of objects in digital images, including animals, products and even the faces of your friends.

We can analyse a trained CNN model to see the technique that it has learnt at each of the filtering layers. What we see is that this artificial neural network is doing something very similar to the biological neural network in animals. The training process allows the CNN to learn the general features of cats from lots of different pictures, taken from many different angles and in lots of different lighting conditions. A CNN system can eventually learn to generalize the features that will allow the model to reliably recognize any cat.

To do this well takes many thousands of images and perhaps hundreds of thousands or even millions of iterations forwards and backwards through the CNN model, but this complex training stage can be automated to run on a set of computers. The machine is learning from the labelled training information using a special type of computer program that describes the way in which the artificial neurons in the deep learning system should be connected. This AI method is not telling the machine what to look for, it just describes how it should filter and pool the data. By providing lots of labelled training information, and by using this label to judge what the deep neural network is looking at, the system will eventually learn a set of hidden parameters that will allow the system to recognize these objects. These special AI programs are surprisingly simple and can be written with as little as forty lines of text.

Since 2012, when the first useful systems emerged, CNNs have been rapidly improved by adding new methods that are also inspired by other structures found in the brain. For example, researchers have copied from pyramid cells found in the cerebral cortex to create what we call 'residual networks' (or ResNets). These provide additional paths that skip some layers to build in shortcuts. This has been found to improve the accuracy of the systems. ResNet models can now sometimes detect objects in images with higher accuracy than a human.* And researchers are making new breakthroughs all the time.

Today these AI vision systems are being used to recognize your friends in images that you post on social media; they are used in manufacturing production lines to recognize defects; they are used to assess photographs from a car accident to then produce your insurance claim form automatically; and in driver assistance systems, to recognize people, cars, speed signs and objects on the road ahead.

However, to support the huge amount of computing power that artificial intelligence requires, we needed to develop something much better than the early electronic-valve-powered machines. The most important of all our modern technologies is hidden away out of sight but it's the technology which powers our modern world.

---

* A different filtering approach, called 'group convolutions', has also been used to build even more accurate AI vision systems known as EfficientNets.

5

THE TECHNOLOGY THAT BUILT AI – PART 2: SEMICONDUCTORS, SOFTWARE AND ATTENTION

INTEGRATED CIRCUITS THAT WENT TO THE MOON

On 20 July 1969, when Neil Armstrong made 'one small step for man, one giant leap for mankind', one of the key reasons he had got there, on the Moon, was thanks to a technology that had been developed at 844 East Charleston Road, Palo Alto, California – in what we now call Silicon Valley. The building is still there, and you could easily drive right by. If you do take the time to stop and look, you will find a plaque by the door that reads:

> At this site in 1959, Dr Robert Noyce of Fairchild Semiconductor Corporation invented the first integrated circuit that could be produced commercially, based on "Planar" technology.

The work that went on in this building and the impact that it has had on our lives over the last sixty-plus years is profound. Hidden away inside all our electronic devices, and invisible to most people, semiconductor integrated circuits – also called ICs and more commonly 'microchips' – are perhaps the single most important technology device that humans have created. They are a true miracle of engineering innovation.

So fundamental are semiconductor integrated circuits that I believe their development may explain one of the great mysteries of our universe. The story goes that Enrico Fermi, considered one of the architects of the nuclear age, was talking over lunch at the New Mexico Los Alamos labs in the summer of 1950 with some fellow physicists, discussing the fact that with over 100 billion galaxies in the universe, it is mathematically highly probable that there are other intelligent beings out there. And then he asked: 'But where are they?', coining what we now call Fermi's Paradox to describe this discrepancy. I think the reason that we have never met these other intelligent beings may be because they haven't built semiconductor integrated circuits. Their advanced societies are stuck in the 1950s. They haven't been to their Moon, they've never built an internet, they have no small, high-performance computers, no advanced telecoms, no smartphones, no deep-space probes beaming back pictures and no artificial-intelligence technology that helps them solve fundamental problems. All because they didn't invent transistors.

Transistors are switches, and are similar to valves in the way they work, but instead use semiconducting materials. They were invented in 1947 at Bell Labs in New Jersey by John Bardeen and Walter Brattain together with William Shockley. All three shared the Nobel Prize in Physics in 1956 for their invention of this incredible device. By applying a voltage to a control pin (called the 'gate') that sits in the middle of the transistor, the semiconducting material can be turned on or off. Like the valves used in Colossus, transistors can be used to build computers, but they are much smaller and need much less power.

A transistor is an amazingly advanced piece of technology, and it is completely possible that we would have continued to gradually improve each transistor and ended up using lots of them to build better electronic systems. Maybe our electronic computers would have been 100 times or even 1,000 times better than the original computers, built with valves. However, if we had not discovered a way of integrating lots of transistors together in a single chip, we would never have been able to make the progress that has been achieved over the last sixty years. Instead of having just one transistor per silicon chip, we can now fit over 100 billion transistors inside a single integrated circuit, which is just a bit bigger than your thumbnail. The rate of technological progress since the first IC built in 1960 is unprecedented. Modern integrated-circuit semiconductor devices are perhaps the most advanced products that humans make.

It was electronics engineer Robert Noyce and a team at the semiconductor company Fairchild Semiconductor who built the first 'monolithic integrated circuit', which combined a number of transistors all working together on a single piece of silicon, rather than using multiple separate pieces. It was this breakthrough that has allowed us to scale up our technology as fast as we have. On 14 January 1959, an electronics engineer in the Fairchild team, Jean Hoerni, helped refine this into a 'planar semiconductor process', allowing them to connect transistors and make them work together in a single integrated circuit. Based on this technology, the first monolithic integrated circuits were built at 844 East Charleston Road in September 1960. These first ICs integrated just four transistors to build a simple logic circuit on a single piece of silicon, which could be used as a building block for creating smaller computers.

Developing this planar manufacturing process was complex, and the first devices were extremely expensive, selling at over $1,000 each. The whole project was nearly shut down because no one could see who would pay for these very early microchips, when you could buy individual transistors much more cheaply and just connect them together yourself. However, one very important customer stepped forward who needed the very smallest and lowest-powered electronic devices.

In 1961 President Kennedy had announced that he wanted the USA 'to land a man on the Moon and return him safely to the Earth'. The Apollo programme was by far the largest customer for Fairchild's early integrated circuits, and without this driving project and the huge budgets that were involved it is possible that semiconductor ICs would never have progressed. During 1963 the Apollo project consumed more than 60 per cent of all the integrated circuits manufactured. Without the small size and reduced power that these integrated circuits provided, it would have been impossible to build the guidance computers that could take Armstrong, Aldrin and Collins to the Moon.

*

In 1965 another electronics engineer in the Fairchild Semiconductor team, Gordon Moore, wrote an article for the thirty-fifth-anniversary issue of *Electronics* magazine that he called 'Cramming more components onto integrated circuits'.[1] In this article, he provided a prediction for how integrated circuit technology might develop and stated that he thought the semiconductor industry could find ways to double the number of transistors on a chip every year for the next ten years. This implied a massive 1,000-fold improvement, which seemed unbelievable at the time. By 1975, this goal had been achieved and the term Moore's law was being used to describe semiconductor technology progress. At that point, Moore (who together with Bob Noyce and Andy Grove had founded Intel in 1968) updated his prediction, saying that the rate of progress would now slow, doubling the number of transistors every two years over the next decade. By the early 1990s the industry had increased the number of transistors by a further 1,000-fold, and by 2020 a further 1,000-fold increase had become possible. This compounding growth has driven a 25-billion-fold increase in the number of transistors on a single chip since the first IC devices were built in 1960. If your car had improved this much, you would now be able to easily travel at around 200 times the speed of light.

Alongside this reduction in transistor size, power consumption per transistor fell massively, meaning that electronic devices have since become much more powerful and much smaller, and also use much less power. Computers have shrunk from being the size of a room to something that fits in your pocket, while the performance has massively increased. Your smartphone can manipulate photos and videos in a way that until recently was only possible at a big-screen movie production company. It allows you to play realistic-looking 3D games and can store massive amounts of information. Integrated circuits power all the other wireless communications technology that allows you to connect with anyone anywhere in the world and to access information, shops and banks directly from your device.

Advances in semiconductor technology have unlocked thousands of breakthroughs in science and technology and are so woven into the fabric of our modern lives that most people don't even realize they are there. It is the most complex technology that society has ever built, and we often take it completely for granted. Without semiconductors, the Information Age that we now live in would still be science fiction.

Semiconductors have allowed us to start exploring the next frontier of intelligent machines. However, as I discovered when I built my first computer, semiconductor-powered machines also need software. The development of software is a story of innovation and technological breakthroughs that in the early years were led by some amazingly talented people. This progress starts as far back as the Victorian era, with the very first mechanical computing machines and the Countess Ada Lovelace.

THE SOFTWARE PIONEER

Ada Lovelace was born in 1815, the daughter of the poet Lord Byron and mathematician Anne Milbanke (Lady Byron). Byron left his wife a month after Ada was born and her mother, concerned that she might follow the wayward ways of her poet father, encouraged her to study mathematics instead. In June 1833 Lovelace met Charles Babbage, who at the time was Lucasian Professor of Mathematics at Cambridge University. This prestigious post had previously been held by Isaac Newton and would later be taken by Stephen Hawking.

Babbage was developing his 'Difference Engine', a mechanical calculating device, and in 1833 Ada Lovelace described seeing a small prototype machine: 'We went to see the thinking machine (or so it seems) last Monday. It raised several numbers to the 2nd and 3rd powers and extracted the root of a Quadratic equation.'[2]

Babbage moved on to develop his Analytics Engine, a more complex general-purpose mechanical computer, and in 1840 gave a seminar in Switzerland describing it, from which a journalist wrote an article in French. Babbage sought help from Lovelace to translate the article into English, and while doing so she added a great many of her own notes. These notes[3] include what many consider to be the very first computer program. Her step-by-step sequence of operations describes a software algorithm that could be run on the Analytics Engine to calculate Bernoulli numbers, which are important in probability theory. However, the machine was never completed, and so the program was never run.

Lovelace also described the concept of a high-level computer-programming language, saying, '. . . a powerful language is developed for the future use of analysis, in which to wield its truths so that these may become of more speedy and accurate practical application for the purposes of mankind'.[4] Ada Lovelace was well ahead of her time and showed incredible insight. Unfortunately, she died from cancer at the age of just thirty-six. On the second Tuesday of October each year, we should all remember her achievements and those of other female engineers as we celebrate Ada Lovelace Day.

Lovelace's insight – that a sequence of simple instructions could build and describe a complex mathematical process – was built upon by Alan Turing. It is said that Turing was heavily influenced by Ada

Lovelace's description of software and that this helped him to develop his concept of a 'universal machine', or what we now call the Turing machine. Turing showed that even the most complex tasks could be broken down into a long list of simple instructions acting on data. As we added memory capacity and compute-performance to build modern computers, so software and the functions that it describes have become incredibly powerful.

Driven by semiconductor-powered computers and advanced software, the frequency of major breakthroughs in AI has been increasing. Within perhaps the last ten years, the most significant have been those focussing on memory and attention.

MEMORY AND ATTENTION

Endel Tulving with his cat, Cashew

Psychologist and cognitive neuroscientist Endel Tulving said that his cat Cashew always appeared to know when it was going to rain – she would run for cover every time just before the clouds broke. But he also noted that despite Cashew's obvious intelligence, she never thought to carry an umbrella. 'Animals are quite happy to do the Darwinian thing and adapt to the environment as it exists,' said Tulving.[5]

Tulving was one of the leading figures researching how humans retain information. We are able to memorize and, more importantly, generalize information that we find important. Our brain can focus attention on key words, on parts of an image, or on other relevant details. As the brain realizes this information will be required over a longer time span, it appears to generalize it even more and will find different ways to retain it.

Tulving developed the idea of a 'multi-store-model'[6] to try to describe how our long-term information storage system works. In this he suggests that humans store and recall long-term memory information through a layered process, as follows:

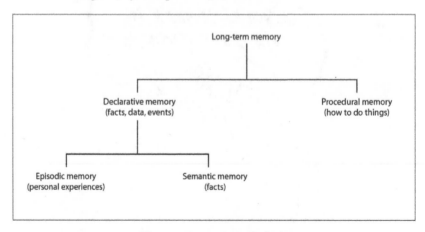

How memory works in the brain

In the first level, the brain makes a distinction between:

- remembering how to do things (what he called 'procedural'); and
- remembering meaningful events (what he called 'declarative')

Examples of procedural memories include knowing how to tie shoelaces, write with a pen, type on a keyboard or eat with a knife and fork. Your procedural system remembers the set of actions that allow you to perform tasks even after you have forgotten when you learnt them. These procedural actions quickly become subconscious, and the method comes back to you even many years later, like riding a bicycle.

I remember when my boys were younger and they were trying to learn to ride a bike. With a bit of help they managed to get going and then wobbled off down the lane. The pedalling was going well, and speed was building. The balance was not too bad, but they were still struggling to remain upright, which made it hard to make the bicycle go straight. I remember one unfortunately positioned bed of stinging nettles at the side of the lane that managed to catch one attempt – the subsequent fall was itchy and a bit painful. This procedural-memory-training attempt for our boys is now among my 'declarative' memories.

Your declarative memory system is about these kinds of events. It allows you to remember a highlight from a family holiday (even if you have forgotten exactly when and where this took place) or an important event during your schooldays or at work. The declarative system breaks down into two connected systems that help you remember events and remember relationships, which Tulving respectively called 'episodic' and 'semantic'.

As an example of an episodic memory, whenever I smell gingerbread, I am always reminded of my childhood. My mother would often bake a loaf of gingerbread at the weekend and then leave it to cool in our family kitchen. When I now smell gingerbread, this will trigger a clear picture of the room and spark other memories too. For instance, our family dog Cider once managed to steal a whole gingerbread loaf that had recently come out of the oven. She ate it in three big bites and the poor dog was very unwell. Most likely she would have had much less fond memories from the smell of gingerbread.

This episodic memory retains information relating to events that have happened to you and also includes information about locations that can become quite specific – what Tulving called 'temporally dated episodes'. Your episodic memories appear to be linked to your five key senses, which is why they can be triggered by a sight, a sound, by taste, touch or smell.

Your semantic system is different. For me, it includes the name of my sister, the fact that she lives in Toronto, that Toronto is in Ontario, Canada, and that Toronto is also not too far from the spectacular Niagara Falls. The semantic system is a memory of relationships (a bit like a mental thesaurus) and describes how things relate to each other. It also contains facts. In addition, your semantic memory is needed to understand the broader context in language – for example, knowing that July is the month that follows June, or that Wimbledon is both an area of London and the name of a Grand Slam tennis tournament.

When you first learnt words, you used your episodic memory, but as this information is retained it becomes part of your semantic store and the two systems work closely together to provide generalized recall. This layered memory system is perhaps why stories appeal to us, as they allow us to recall information using our own experiences. I always find that a well-written novel will draw images in my mind that are far more powerful than any movie could show. As a result, I always find it slightly disappointing when I see the movie of a book that I enjoyed. It never quite lives up to the pictures that I had painted in my own mind. If instead I read the book after watching the movie, it is never quite the same immersive book-reading experience.

Memories in our brains are not filed away neatly in some kind of indexed filing cabinet. You don't need an index code or a special address to find a relevant memory. Your brain appears to be able to instantly recall information that is related to your current stream of thought or your current activity. Your recall is associative, like a mind map, and relevant information is found hidden in the connections between neurons in your brain. It is even built into neurons and synapses as sequences that perform muscle movements that you can trigger from a single impulse.

This concept – that our memories are linked to our thoughts and that memories are not stored away but are instead hidden in the connections between our neurons – takes some imagining. For people you know well, you will have very precise memories captured by the neurons in your brain. But you will probably have wasted a few neurons storing some knowledge even about people that feature regularly in the news. You may not like the idea of this, but you very likely have some specific Donald Trump-, Boris Johnson- and Jeremy Corbyn-related neurons and synapses in your brain.

Linked to memory is language, which is one of the ways in which we share information and knowledge. Babies don't understand language when they are born. This is a skill that all humans must learn. We can easily guess that this is a learnt skill because as Homo sapiens has spread out across the planet, different geographic groups have developed their own languages, their own dialects, and even their own regional accents. Humans' understanding of language is passed down from one generation to the next through education.

Language can be extremely complicated and sometimes very confusing. The English language, for example, includes words and phrases from Ængle, from Norse, from Latin, from Saxon and from Norman (itself a mix of Norse, Latin and Frankish), and has more recently added many slang words from dialects all over the Anglophone world. As a result, English contains many words that have a double meaning, and objects can often be described with two different words (the Anglo-Saxons farmed 'sheep' so that their Norman masters could eat 'mutton').

Even between two people who both speak English, it is sometimes difficult to follow the conversation. As an example, in September of 2021 I was in a black taxi in London and the driver said:

'Did you hear about Ronaldo? They'll play much better now, don't you think? I heard that it was the governor that made it happen. City must be really upset."*

To translate, you first need to be able to decipher that the general context of the conversation is about football (or soccer in American English). You also need to know that 'Ronaldo' is referring to Cristiano Ronaldo, a famous football player who had just signed for the English Premier League club Manchester United from the Italian club Juventus. 'The governor' refers to Sir Alex Ferguson, the very successful former manager (coach in American English) of Manchester United. Ronaldo had played for Manchester United earlier in his career, when Sir Alex Ferguson was in charge, with the two men building very strong mutual respect. 'City', as you may have worked out by now, refers to Manchester City, the bitter local rivals of Manchester United. They had won the English football Premier League the previous year and it was rumoured that they, too, had tried to attract Ronaldo.

What we can see from the taxi driver's comment is the challenge that a machine might have in trying to understand language. The meaning of words is often dependent on their context. Even trying to understand the differences between English and American English is challenging. A machine will need to understand vocabulary, sentence construction, language semantics and context within sentences. But it will also need to have an understanding of the wider subject matter, operating over different time spans.

This means that an artificial intelligence system also needs a memory system. This will allow it to retain information and then use it later as it reads through a text. However, the memory system cannot store everything – that would be inefficient. It needs to be able to focus attention on the key words and phrases that might be needed later. This leads us to the very important concept of an 'attention system',

---

* A slightly different phrase was used here, but the meaning was the same.

which allows AI to understand and remember the key items and to establish context. Adding an 'attention system' to a machine learning method greatly increases its capability.

Early work by AI researcher David Rumelhart in this area led to a method for building memory into AI using something called a 'recurrent neural network' (or RNN). The idea here is that information is held in a short-term memory and as the subject 'recurs' the AI system works out which information is important. The early recurrent neural network AI systems then led to something called 'long short-term memory' (or LSTM) systems, which were developed by AI researchers Sepp Hochreiter and Jürgen Schmidhuber back in 1997.

LSTMs allow key words to be stored and then retrieved later by the machine learning system. This lets the machine learning method capture context from a sequence of words that make up a sentence or a paragraph. The LSTM can use the context that it stores to understand subsequent words in the sequence. For example, if I tell you that 'my cat is called Ginger, and it is very fluffy', you can understand that the 'it' I am referring to is my cat. From my conversation with the taxi driver, we know that the 'they' the taxi driver refers to is Manchester United. It comes as second nature to us, but without LSTMs it would have been hard to develop AI systems that could start to understand language.

They have also been used to translate one language to another. Often the order of words in one language, such as English, is different when compared to other languages. For an AI system to translate a sentence from one language to another, it needs to store the words, translate them and then reorder them to match the new language.

These RNN and LSTM developments are useful, but it was only as recently as December 2017 that a big breakthrough was made. A now-famous technical paper titled 'Attention is all you need'[7] was published by a team from the Google Brain project and Toronto University. This much-referenced paper proposed an improved mechanism for learning how to focus 'attention' on the key words in sentences and paragraphs. The paper defined a process it called 'multi-headed attention'. The idea here is that rather than looking at the words in a strict sequence, you chop the sequence up into the individual words, adding an identifier to show where they sat in the original sequence of words. You then try to learn the importance of each of these words in parallel and learn how they relate to the other words in the sequence. The idea was to build a semantic under-standing of language and work out which words need most attention.

This machine learning method has been dubbed a 'transformer' (the GPT in ChatGPT stands for generative pre-trained transformer), and since 2018 this AI training method has been extremely successful in many different natural language processing applications, including translating between different foreign languages, performing sentiment analysis on customer service replies to understand what people are feeling, and for generating text built from a 'seed' of information. You can find many examples online for text generated by an AI system trained with the GPT-4 transformer model built by the AI research firm OpenAI. The company's ChatGPT language system has caused quite a stir and highlights how powerful these systems are becoming. You can now find examples of children's bedtime stories created from just a short set of prompts that describe the subject, software code that has been automatically generated from a short logical description and sonnets written in the style of Shakespeare, all produced by this extremely powerful transformer-based AI system.

Transformer AI models are being applied today in many different types of language system. As an example, when you use a regular search engine, the search term that you enter is fed to a powerful transformer-based natural-language AI model. The system tries to make sense of what you are asking. It will spot spelling mistakes, check whether you actually meant the specific phrase entered, and may sug-gest similar phrases that might be more accurate. The AI system can also translate foreign languages to help you find alternative content.

You may have noticed that in the past if you entered a longer search term, the search engine seemed to become confused. Transformer-based AI systems now help the search engine to make sense of these longer sequences. Try entering a complete sentence and the transformer AI system will now search through this long sequence of words to extract the key words that it should focus attention on. Today a search engine is much closer to a question-and-answer system and this capability is quickly growing better as the next generation of transformer-based natural language AI systems becomes available.

The concept of multi-headed attention is not only being applied to language but is now also being used for other sequence problems, such as understanding how proteins fold. Proteins are found in each of your cells and perform a vast array of functions in your body. They are made up from a long sequence of amino acids, with millions of different types of proteins each being expressed by a different sequence. These sequences build up into extremely complex three-dimensional structures, which then allow the protein to take on different biological functions.

Trying to understand the incredibly complex protein sequence, and how it folds, is a fundamental challenge in science, which will unlock new breakthroughs in drug discovery and healthcare. AI can show us how each protein folds, allowing scientists to understand the function that it can perform, and lets us learn how to use this protein to perform a specific function at the cellular level. For example, you could connect a drug molecule to the protein in a cancer cell in just the right way for the drug to stop the cancer from growing but without any side effects. Recent breakthroughs in transformer-based AI systems that have been created for understanding how proteins fold have produced incredible results.[8]

As one example, the DeepMind AlphaFold system is able to predict the structure of a protein from its amino acid sequence. Traditional experimental approaches had previously captured around 190,000 protein structures;[9] in July 2021 the AlphaFold system was able to produce a library of 350,000 protein structure predictions that included all proteins expressed by the human body. Then in July 2022, in partnership with the European Molecular Biology Laboratory (EMBL) at the European Bioinformatics Institute, a stunning scientific achievement was made with the release of a snapshot of nearly every existing protein on Earth, totalling 200 million. This has the potential to dramatically increase our understanding of biology. Even taking into account the limitations, the rate of progress of AI's use in biology in the last few years has been enormous and clearly shows the revolutionary effect of transformer-based AI models on science and other fields.

There is another enormous advantage that this concept of transformers has brought to AI. Transformer AI systems don't need any special labelled information for training. You can take any sequence of words that form language and use these to teach the model how language is constructed. You can just use language information found on the internet, such as text from Wikipedia. This idea of training a machine learning system using unlabelled information is called 'unsupervised' learning, or more correctly 'self-supervised learning'. This pre-training process creates a foundational model that already has a deep understanding, which you can then fine-tune so that it will be better at specific tasks.

The concept of multi-headed attention and self-supervised learning (together described as 'multi-headed self-attention') has unlocked the ability to create very large AI systems. Researchers are now building models with billions and even trillions of parameters, and these have proven to be highly effective. However, as a result the amount of computing power required has also been accelerating. OpenAI published a paper that showed that since 2012 the amount of computing power required to train the largest AI models has been increasing exponentially and is doubling every 3.4 months. This has resulted in the computing power required to train the largest models increasing by 300,000-fold[10] between 2012 and 2019. The computing costs to deploy the ChatGPT system, for instance, are enormous. At the time of writing, the computing systems required to run the system are estimated to cost over $700,000 per day, and as it becomes more and more popular this cost will multiply, making transformer-based AI chat systems much more expensive for the providers of these services than current search engines are.

Transformers have had an incredible impact in the world of AI systems. Following their breakthrough, progress on AI systems for language understanding, for translation and for text generation has been phenomenal. But for all the hype, these systems are not perfect. In fact, they are still a long way behind the incredible ability of humans. The learning method they use is still very simple. It takes massive amounts of information to train the foundational large language models that this transformer approach has enabled. Although the basic system can be trained using any language information, to make the large language models work accurately in specific applications they also need to be fine-tuned on information specific to the subject area. For example, if you want to create an AI customer service engine, you will need to train the system with all the information about your products and processes, plus the language that might be very specific to your business. As this information changes, you will need to fine-tune the system again with the new information. You may also become worried about which company you are sharing this critical business information with.

AI systems could be trained to understand specific aspects of the legal system if they are given access to this detailed information. They will then help lawyers review or draft new legal documents. Systems like this are already starting to become available. However, just like a lawyer, this information will need to be constantly updated so that the system has the latest information and all the new legal precedents.

Enormous amounts of compute are required to train these massive AI models, and deploying them is very expensive. New breakthroughs are required to make them more efficient and better able to learn from new information, and more work is still needed to develop even better language models and to further improve this multi-headed attention learning method.

The progress in AI vision systems and AI language systems has been incredible. But developing these amazing AI breakthroughs only became possible because of the work of earlier innovators. Researchers in AI today rely on the earlier achievements of many extraordinary heroes in software development, such as Grace Hopper.

GRACE HOPPER, SOFTWARE LANGUAGES AND BUGS

Grace Hopper was a mathematician who, like Ada Lovelace a century before her, believed that it should be possible to create a programming language based on English, rather than on the low-level logic descriptions that computers use.

In 1949, Hopper joined the Eckert–Mauchly Computer Corporation as a senior mathematician working on one of the world's first commercial electronic computers, the UNIVAC 1. 'It's much easier for most people to write an English statement than it is to use symbols,' Hopper said, in 1954. 'So I decided [programmers] ought to be able to write their programs in English, and the computers would translate them into machine code.'[11]

She called this software technology breakthrough a 'compiler'. It took her three years to convince the company to accept her idea. The program that she developed was known as the B-0, and she published her first paper on the subject in 1952,[12] along with her 'English-style' programming language FLOW-MATIC.

In 1959 Hopper served as a technical consultant for the US government and industry consortium Conference on Data Systems Languages. It was here that she defined an even better programming language – her COBOL language went on to become the standard business programming language for computing over the next decades.

Grace Hopper was known as 'Amazing Grace' by her colleagues in the computer industry and was famous for her entertaining talks and visual descriptions. To show why computers must be small, she would hand out a 30cm-long piece of wire and explain: 'This is a nanosecond' – the time it would take for an electric current to travel down this short piece of wire. She would then highlight that, with a 300m length of wire, the journey would take one millisecond. She was illustrating that the reason fast computers must be small was not a matter of coding, but a hard limit imposed by the physical world.[13]

There is one other important part of software development that Grace Hopper is also responsible for discovering. Taped inside one of her notebooks is a moth or 'bug' that she found on 9 September 1947. This 'first actual case of [a] bug being found' was discovered by Hopper when trying to find the cause of an error in a program running on the electro-mechanical Harvard 1 machine that she was working on at the time. On opening the machine, she and her colleagues were surprised to find the insect trapped in a relay and they had to 'de-bug' the computer to fix the problem. In Hopper's later software work, whenever she discovered a problem, she would call this a 'bug' and then work to 'de-bug' her software code. Her now-famous notebook is held at the Smithsonian museum, with the bug and the tape, carefully preserved as a piece of unique history.

In 2016 Hopper was posthumously awarded the US's Presidential Medal of Freedom, for her pioneering work in the field of computer science, by President Barack Obama. But Amazing Grace was just one of many women who contributed to the growth of software. In the early days of computing, building hardware was considered the difficult job, whereas telling the machines what to do was seen as laborious and less important, like the job of a human computer (that is, a mathematician who solves equations), and this role was often given to women. Kay McNulty, Betty Jennings, Betty Snyder, Marlyn Meltzer, Fran Bilas and Ruth Lichterman were all top mathematicians who had worked as human computers at the Moore School of Electrical Engineering at the University of Pennsylvania. They were the first computer programmers to work on one of the very earliest electronic computers, named ENIAC, built in 1946.

As a result of the work that these women did, ENIAC was able to run missile-ballistics trajectory calculations that relied on complex differential calculus. However, when ENIAC was revealed to the public, the women and the work they had done received no recognition. Throughout the 1950s and 1960s the story remained the same. Women who had worked as human computers were asked to help program the new electronic computers. They were often the only people able to get these machines to do anything useful and yet, too often, the work of these unsung heroes went unrecognized until much later. One woman who deserves much more recognition is the one who helped put a man on the Moon.

MARGARET HAMILTON AND SOFTWARE ENGINEERING

The Guidance Computer that allowed Apollo 11 to reach the Moon was programmed by a group from the Massachusetts Institute of Technology (MIT) led by a brilliant software engineer called Margaret Hamilton. She and her team drove the development of many concepts that are now commonplace in modern software. The software built into the Guidance Computer was designed to deal with complex multi-faceted problems that needed to be constantly monitored and controlled. This was made possible by a new concept of switching between different software tasks. Today we call this software technique real-time multitasking.

When the Apollo Lunar Module was just three minutes out from the surface of the Moon, the computer began flashing a warning message. Mission Control back in Houston analysed the error codes and, thanks to the confidence that they had in Hamilton's real-time control software, they were able to tell the astronauts to 'go for landing'.

If you watch a video of the Apollo 11 Moon landing, you will also see a point where Neil Armstrong becomes concerned about the rocky surface that the Lunar Module is heading towards, and he says, 'Switching to manual.' The joystick that he then uses to steer the craft is in fact still an electronic 'fly-by-wire' system: his inputs were being interpreted by the Guidance Computer and the activation of the rocket thrusters was still controlled by the software.

Hamilton is credited with coining the term 'software engineering', and on 22 November 2016, at the same ceremony where the posthumous award for Grace Hopper was presented, Hamilton was awarded the Presidential Medal of Freedom for her own pioneering work.

Software has gone on to become a dominant technology over the last thirty years, and this is perhaps best summed up by a 2011 blog post by venture capitalist and founder of the original internet browser company, Netscape, Marc Andreessen, entitled 'Why software is eating the world',[14] which highlights how software allows rapid innovation and is disrupting traditional industries. We should never forget that this critical field was pioneered by some of the most brilliant women, who between them developed this new field of software engineering.

*

The ability to see, or to understand language, is a hugely complex process, but it is the breakthroughs in computers, semiconductors and software that have allowed us to show a machine how it can learn to perform these functions. But can this same approach let a machine do things that we think of as being uniquely human, such as composing pictures, coming up with creative ideas, or finding obscure connections between different pieces of information?

6

THE TECHNOLOGY THAT BUILT AI – PART 3: GETTING CREATIVE, GETTING CONNECTED, AND GETTING MORE INFORMATION

An object that has some legs, a flat base and a back rest is probably a chair. The specific object that you are looking at may be a rather strange designer chair but if it has these basic components, you will quickly recognize it as a chair. Humans are very good at recognizing new objects and figuring out what they are for. Traditionally, when AI systems see something new that they have not been trained to recognize, they may struggle to categorize this new object in the same way a human could. But a new AI method called 'one-shot learning'[1] is starting to solve this problem.

The Web is now full of images that have captions associated with them. People post pictures together with text that might read something like 'My cat is sitting on my laptop again.' The Web is also full of pictures of furniture, natural environments, mountains, towns, cars and animals, and will usually have useful labels associated with them. By a combination of an image-recognition system to recognize the features from the pictures and a natural language system to read the labels, new AI systems can match the features that they find in the picture to the words that they can understand from the captions.

The picture of the cat sitting on your laptop keyboard helps the

AI system learn what cats and laptops look like. By finding associations between these two information sources, the system will start to recognize what objects are in the images. Once trained, the system has understanding about both the objects and captions. When presented with a new object, it can use its knowledge model, which has been previously trained, to work out what caption might be appropriate for this new object. This AI learning method allows one-shot AI systems to quickly recognize new objects by finding the closest match from its knowledge model.

An AI-generated picture of a cat

Recently, a new AI method called 'diffusion models'[2] has started to emerge as a way to generate photo-realistic images. These work by taking labelled images and then slowly adding corruptions that wipe out details in the image until the picture becomes just random pixels. A neural network is then trained to reverse this corruption process so that the original image can be recreated. This training process then allows the resulting trained diffusion model to generate new photo-realistic images just by being given a simple text prompt.

By joining together this diffusion model approach with a one-shot-learning approach, a text caption such as 'Create a photo of a Persian

cat, wearing a cowboy hat and a red shirt, playing a guitar, on a beach'
will create a photo-realistic picture from this text. If you prefer, you
can ask it to produce the image as a cartoon, or in the style of a
famous artist.

An AI-generated picture of a cowboy guitarist cat

So-called 'generative' AI has the potential to completely change the
creative industries. Today a huge amount of effort goes into laboriously rendering animations and backgrounds for computer games and
other media. These generative systems will accelerate this work. In
music, generative AI systems are also starting to provide new ways of
producing music for backing tracks. In the future, much more effort
can be placed by humans on the creative input rather than on the
laborious production work, and scenes or music that don't work in an
animated film or computer game could quickly be changed without
wasting huge amounts of effort or expense.

Generative AI will certainly not replace human creativity – in fact,
writing prompts for these generative AI systems is becoming an artform
in itself – but it will help to create new forms of art and increase productivity, allowing much more content to be produced, much more quickly.

AI also brings the prospect of much more customization. An AI system that helps people to develop their own computer games could easily become possible. These games could then be shared using social media networks and others could easily change them or add to them. The wisdom of the crowd may end up creating much better games and entertainment content, and could unlock a whole new era of creativity.

But there are challenges, too. In April 2023 a TikTok social media user going by the name of Ghostwriter977 took the internet by storm using artificial intelligence to clone the voices of Canadian rapper Drake and singer-songwriter Abel Tesfaye, better known as The Weeknd. Using prompts and generative AI systems, this person produced a song titled 'Heart On My Sleeve'. The track went viral, with users on Twitter calling the song 'scary' because it was so realistic. But Spotify and other music and social media sites quickly removed it from their platforms after Universal Music Group stepped in and asked for it to be taken down, with a spokesperson telling *Billboard* magazine that such posts 'demonstrate why platforms have a fundamental legal and ethical responsibility to prevent the use of their services in ways that harm artists'.

Although it is easy to see how generative systems can be useful, we also need to make sure that they are not used to generate deepfake material. Copyright issues and ownership rights need to be addressed. We will look at these challenges and the controls that are needed a little later. First, we need to understand how we are able to make sense of connected pieces of information in the first place.

FINDING ASSOCIATIONS IN INFORMATION

When my children were younger, we would sometimes play a word-association game that would go something like this:

Child: 'London'
Parent: 'Paddington' (a railway station in London)

> Child: 'Marmalade' (the children's character Paddington Bear's favourite food)
>
> Parent: 'Dundee' (a town in Scotland where marmalade was originally made)
>
> Child: 'Grand Theft Auto' (a popular computer game that was developed by a company in Dundee)

Each of the 86 billion neurons in your brain connects, on average, to around 10,000 other neurons. Your brain's ability to create connections and find associations between seemingly random sets of information is amazing, and far better than today's machines.

When you hear a sound, or pick up a particular smell, you are kicking off connections in the brain that will find lots of other associated information from your prior knowledge, from your understanding of the world, and from your memories. One trick that 'memory experts' use for remembering people's names is to anchor the information with additional associations so that they can connect the name of a person with something memorable, such as 'Carol – who spends the summer in Cornwall'. Adding such associations gives you other routes for remembering information. Making the associations visual, or building them into a story, helps even more.

What your brain is doing is searching in what is called a 'high-dimensional space' to find connections. We live in a three-dimensional world, and we travel through time, which adds a fourth dimension. It's hard for humans to visualize a high-dimensional space, but these do exist, and your brain deals with them every time you make a complex association.

One simple way to think of a high-dimensional space is as an extremely large grid made up of groups of boxes. Some boxes hold information and others are empty. When you find something interesting or relevant in one box, you can look in the neighbouring boxes to see if there are any associated pieces of information. The problem with our grid analogy is that it doesn't show the complex interconnectedness of possible associations. A high-dimensional space is actually more like a three-dimensional web, with links between all the connected items and where each link can have a number that describes how closely related these pieces of information are. In fact, this looks a lot like your brain's connectome, the wiring diagram representing the massive set of connections between neurons in your brain.

In mathematics there is a whole subject area related to this called 'graph theory'. You may be more familiar with simple line and bar graphs, which typically show the relationship between two quantities. But there are examples of much more complex graphs, which you are perhaps also familiar with. Social media networks are a graph that captures the connections between you and your friends, and even your friends' friends. This allows the social media platform to work out how everyone is connected. From this they can try to decide which of your social media posts other people might be interested in.

The World Wide Web is also a very large and complicated graph. A search engine crawls over the Web, using a tool called a Web spider, and builds a complete graph that captures all these different website connections. Because Web pages, and the links between them, are being added and changed all the time, keeping the search engine up to date is a never-ending task, the Web spiders of major search engine companies continuously crawling all over the Web to maintain this high-dimensional graph of it.

Like a brain, an artificial intelligence deep neural network also makes use of graph theory. In order to understand this subject better, we need to visit the ancient Prussian town of Königsberg (now Kaliningrad in Russia) for a thought experiment.[3] The River Pregel flows right through the centre of old Königsberg, with two large islands in the middle of the river. The old city was divided into four districts, with a district on each riverbank and a district on each of the islands. Connecting all the four areas together was a set of seven bridges. The challenge for the good citizens of Königsberg was:

could they visit all the bars spread out across the four areas by crossing each one of the seven bridges only once? Sounds simple, but however much beer the poor citizens of Königsberg drank, they could never find the solution.

The mathematical challenge is to prove definitively that there is no solution. Just trying again lots of times and not finding an answer doesn't prove this. (*What if there is a solution that just a bit more beer might uncover . . . ?*)

Eventually, in 1736, the Swiss mathematician Leonhard Euler offered a proof (with no beer involved; not surprisingly, given that he was described by one of his patrons as 'too sober'). In doing so, he developed the mathematics of graph theory.[4]

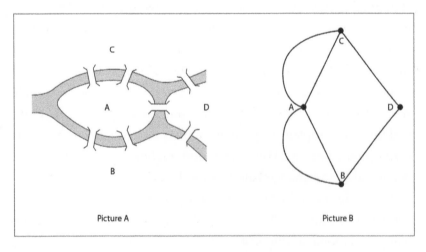

Picture A   Picture B

The Königsberg bridge problem

By turning the map of Königsberg (picture A, above) into a mathematical graph representation (picture B), with four nodes (or dots) each representing the different areas of the city, and then connecting these areas with seven links that represent the bridges, he was able to analyse this more abstract graph and give his proof. The maths is still complex but basically identifies the fact that because all four (an even number) parts of the city have either three or five (both odd numbers) bridges connected to them, at some point you will end up back in the same bar that you just left, or you will need to cross the same bridge twice.

The analysis of these complex graph data-structures is among the new areas in AI that have been very rapidly expanding, with a technology that is called 'graph neural networks' (or GNNs). Researchers have developed neural networks that operate on these graph data-structures, such as the 'complex social-network graphs' that large social media companies need to understand. These GNN AI methods are also being used to identify fake news and to make better recommendations on e-commerce sites. Often, transactions and social interactions are very dynamic – you may be interested in a new book that has just come out, but to find the book the graph must be updated and this addition must be known to the GNN. These types of additions and changes to the graph can be modelled in a special form of graph neural network called a 'temporal graph network'.

Humans are very good at making associations in areas where they have built up experience. AI systems, however, can be focussed on areas where humans might struggle to learn all the complex details. If we have the information to train an AI system, then it can develop a very deep understanding of a subject, such as understanding all the connections in the World Wide Web. The same can apply to complex structures such as the connection of atoms in molecules. Graph neural networks and associated AI approaches are helping humans solve problems that they couldn't easily solve on their own, making this an exciting area of AI research that promises many new breakthroughs.

These systems that can understand complex relationships, such as how molecules and proteins interact, represent an exciting new area in artificial intelligence. Another area that will become extremely important is the development of AI systems that can learn to learn.

LEARNING TO LEARN

Elephants have an unlikely friend in an AI technology called 'reinforcement learning'. Between 2013 and 2016, the Allen Institute, a non-profit organization based in Seattle, ran the 'Great Elephant Census',[5] which had the ambitious goal of counting all the savanna elephants in Africa. These enormous creatures, our largest land animals, are extremely smart and curious, with a prodigious memory, and they build extremely close family ties. They are also critical to maintaining biodiversity and a healthy ecosystem. However, elephants continue to be hunted for their ivory and for meat, and are also sometimes poisoned by farmers who want to keep these large animals off their land. To understand the threat from poachers, researchers from the Allen Institute used aerial surveys and a robust statistical process to count not just living elephants but also carcasses. What the survey found was that Africa's savanna elephant population had decreased by 30 per cent in just seven years. In some regions, elephants were even facing extinction, such as in north-eastern Democratic Republic of the Congo, in northern Cameroon and in south-west Zambia. However, in some places the decline in population had been reversed. In Uganda during the 1970s, the country experienced severe political turmoil under the brutal military dictatorship of Idi Amin. As a result, the elephant population reduced from around 30,000 to fewer than 800. More recently, as a result of conservation projects and increased patrols, the elephant population has recovered to about 5,000, but they remain severely threatened.

To help in this fight to increase elephant numbers across Africa, AI is being used to model poachers' behaviour and to give the rangers, who try to protect the elephants, much greater insight into how the poachers operate. Given their limited resources, the rangers need to organize their efforts and focus their patrols on the places where poachers are most likely to go. With the large patrol area – which covers thousands of square kilometres – divided up on a map into a grid of 1km squares, an AI reinforcement learning system used twelve years of prior information from over 1,000 cases of poaching to predict the chance of a poacher's snare being in a specific square, with the AI system anticipating and learning from the poachers' past behaviour. In one wildlife park, by using reinforcement learning they were able to quickly find over 500 snares when the typical number found had previously been only around 100 over a much longer time period.

Reinforcement learning is a machine learning method based on describing a 'reward' for achieving a desired outcome, and not rewarding (or giving a negative consequence for) the outcomes that fail to deliver the correct answer – a bit like a point score that you might receive if you do well in a game. A reinforcement learning 'agent' is designed that can perceive and interpret its 'environment'. The agent then takes 'actions' and learns, through trial and error, which of these actions creates a positive effect. Often this is based on working in a simulated 'environment' where historical information (such as information about poachers) can be used to predict new behaviours. The environment is in effect the space in which the agent will operate (such as the large grid of squares in the map of the wildlife park). The agent then learns to perform a set of 'actions' that are determined from predictions made against a model (which in turn has been learnt from information about the environment).

Reinforcement learning is a circular process where the actions may produce a reward (finding a poacher's snare), with this outcome telling the agent that its understanding of the environment was correct. By achieving a positive outcome, the system has reinforced the strength of its predictions about the environment. Equally, if it performs some action that does not result in a positive outcome, this still helps the system to learn which actions are less successful. As the understanding of the environment improves, the agent can then make even better predictions, which will result in more effective actions.

Here, the system is trying to learn through experience. One

challenge, however, is that the system will often need to perform a whole sequence of actions before the desired outcome is achieved, so the system must work out which set of actions is helping it make progress towards getting a reward. This is very similar to how humans learn from experience: children will try different ways of getting what they want, and through experience will hopefully learn that screaming and crying or having a tantrum are less effective than behaving well and saying please and thank you.

In nature, we can see reinforcement learning at work by looking at even a very simple organism that is able to direct its movements. A microbe will learn through trial and error in which direction to move to reach a source of glucose that can provide it with energy. Equally, a robot with sensors and the ability to move in any direction can, by sensing its environment, and through trial and error, learn to find its way around your living room. By attaching this to a floor-cleaning system, you end up with an autonomous vacuum cleaner.

In AI, the idea for reinforcement learning stems from work done by computer science professor Marcus Hutter, in Switzerland, who proposed a model for a 'Universal Artificial Intelligence'.[6] The concept was analogous to the 'universal machine' idea proposed by Turing, in which he showed that a machine could perform any computation. Hutter set out a mathematical proof for artificial intelligence and provided a framework for how we could measure the intelligence of machines. Hutter's model describes a formal method for an intelligent machine that can learn from experience. This method can actually be described in a single mathematical equation that fits on one line.[7]

One of Hutter's PhD students, Shane Legg, who worked with Hutter on this Universal Artificial Intelligence project, went on to become a co-founder of the company DeepMind, together with Mustafa Suleyman and Demis Hassabis. The DeepMind Deep Q system, which learnt how to play computer games and that we saw in Chapter 2, used a reinforcement learning approach. The team then famously built a reinforcement learning AI system called AlphaGo, which learnt to play the complex board game Go by using information about lots of human Go games.[8] This system went on to beat the reigning world Go champion, Lee Sedol.

The rules of Go are deceptively simple. The board is a nineteen-by-nineteen grid on to which players each place, in turn, small stones, with one player taking black and the other white. By watching a small number of games, you will quickly be able to understand the end goal, which is to capture your opponent's stones by surrounding them. The challenging part is working out a strategy so that you can win. The game requires careful and patient positioning that will eventually lead to a slow encirclement of your opponent's pieces. With 250 possible different moves on each go and, in a typical game, players taking 150 goes in total, this gives 250^{150} possible moves in a game, which is equal to 10^{360} (or 1 followed by 360 zeros). Having an AI system master this complex game was seen as a major breakthrough.

More impressive still was the follow-up system that DeepMind built in 2017 called AlphaGo Zero.[9] AlphaGo Zero was not told how to play the game, other than being told the basic objectives, and it didn't use any information from human games. It just learnt by playing game after game against itself, using a 'self-supervised' learning method. After training non-stop for forty days, AlphaGo Zero exceeded the capabilities of all previous systems trained with human information. The same approach was then used for AlphaZero,[10] which is a more generalized game-playing system that was told how chess pieces move and the basics of how to capture a piece, and then played game after game against itself until it became an expert. The same approach was used to learn other games. Even without being trained on how to play, other than the basic moves, it quickly achieved superhuman results.

Reinforcement learning will become a fundamental part of advanced AI systems, and it is particularly useful for testing, predicting and controlling. The technology is being used across many applications, including in self-learning factory robots, in nuclear fusion, in autonomous cars, in finance and in healthcare.

Innovations in AI are coming thick and fast. One key future direction is the combination of different deep-learning artificial-intelligence approaches as systems of learning that work together to achieve more general artificial intelligence. Reinforcement learning will be the glue to ensure this mixture of models can also learn from experience.

Marcus Hutter's equation for intelligence includes one tiny caveat, however. The system must be able to fully understand its environment. As we will see later, our world is far too complex even for humans to fully understand, even billions of years into our evolution. So, to be effective, an intelligence machine will need to be focussed and trained in a specific area. Ecology offers a useful metaphor here: a plant that thrives in one climate will often wither when moved some-where different. This is how we should be thinking about AI systems. When they can become an expert in one specific area, they will be able to deliver real value and will also be easier to control. We should not expect a single AI system to become an expert at everything. We must also bear in mind that real intelligence deals not in certainties, but in probability.

PROBABILISTIC SYSTEMS

When it comes to uncertainty and probability, there are two unlikely sparring partners from the 1700s who ended up arguing about this important subject – one a church minister, the other a famous philosopher.

In his book of 1748, *An Enquiry Concerning Human Understanding*,[11] the philosopher David Hume said that if you are relying on past experiences to infer an outcome, then you must believe that the future resembles the past. Humans want to believe this, he said, but just because something happened before doesn't mean that it will happen again. He argued that this causes humans to be driven by their passions and emotions rather than being led by reason. Using the same argument, Hume proposed that humans shouldn't believe in religious miracles.

The church minister Thomas Bayes had studied both mathematics and theology at Edinburgh University. In his spare time, he continued to work on mathematics. His important work on probability,[12] from 1755, is now widely used in a number of subject areas, including sport, science, medicine, engineering, law and philosophy, and is now also being used in artificial intelligence. It is suggested that one of the key reasons for his interest in probability theory was because he was upset by David Hume's claim that miracles are not possible.

Bayes showed that the probability of an event can be related back to our understanding of the prior state and conditions that caused the event. If we don't have much information about what caused it, then we should set a low probability for it happening again. As more information becomes available about the prior state (that is, what caused the event), we can increase the probability.

To apply 'Bayes' theorem', we need to determine what information is missing and then add new information to fill this hole. As an example, if a pharmaceutical company were testing a new drug and they could see that their test group did not include people of different gender and ethnicity, they would know that by adding people from these groups they could improve the accuracy of their results. This means Bayesian techniques can be used to measure both the probability that an AI system is giving a good answer, and to suggest ways to get better answers. This could become an important part of controlling future AI systems.*

---

* Also, in answer to David Hume's statement that we should not believe in miracles, Bayes highlighted that, because we cannot know all the conditions that might have caused this miracle, there is a (small) chance that the miracle did indeed occur. Using his theory, he pointed out that you cannot be certain the miracle didn't in fact happen.

Marcus Hutter, in his description of a Universal Artificial Intelligence, also referred to Bayes, plus a few other mathematicians and philosophers. These include the Greek philosopher Epicurus (341–270 BCE), who said that all events have causes, regardless of whether those causes are known or unknown. The English Franciscan friar William of Ockham (1287–1347) gave rise to the 'Occam's razor' theory, which states that the simplest solution is probably the most likely. As we have seen, Thomas Bayes said that the probability of events increases as we learn more information about what causes them. The mathematician Ray Solomonoff effectively combined all of these ideas in a paper completed in 1964 called 'A Formal Theory of Inductive Inference'.[13] Solomonoff's innovation was to formally show that for any outcome that you might observe, it is the simplest hypothesis that will have the highest probability, and that increasingly complex hypotheses are increasingly less likely. This led to Hutter's idea that a machine learning system must first understand the environment and then learn more about this environment from actions, with the simplest set of actions being the most likely to have a positive effect. Through this iterative process, the machine can learn how to make progress towards a positive outcome in the form of a reward.

Similar to this reinforcement learning approach, 'probabilistic machine learning' applies Bayes' ideas and tries to improve the accuracy of answers. Probabilistic frameworks can work on their own as learning systems, but they are more powerful when combined with other machine learning methods. They can be used to quantify uncertainty, and then work to reduce this uncertainty to deliver more accurate answers. Probabilistic methods could provide a solution for how we can make AI solutions that are more responsible. As an example, you may have noticed that chat AIs have a tendency to state everything confidently, as if what the AI system is saying is a hard fact, even when they are completely wrong or making things up. Probabilistic methods will allow us to build caution into AI systems, helping them to provide clarity on the uncertainty contained in the answers being provided. The AI system could then learn how to deliver more accurate answers.

The combination of computers powered by advanced semiconductors, together with advanced learning methods, has helped to build our technological age, but artificial intelligence also needs one final ingredient: information.

THE GROWTH OF THE WEB

Cave paintings are perhaps the oldest form of stored human information, dating back more than 44,000 years. Some of the earliest examples have been found in caves on the borders of France and Spain, as well as in Indonesia.

Various forms of written information followed, perhaps linked with the transition that humans made to being an agrarian society. The invention of paper in China around 100 CE sparked an information revolution that would ensure China remained the leading global economy up until the start of the Industrial Revolution. China was also first to develop moveable-type printing, with the oldest printed books appearing there around 1040 CE.

In contrast, the time from 500 CE through to 1400 CE in Europe is often called its 'Dark Ages'. This 900-year period, which began with the fall of the Roman Empire and ended with the birth of the Renaissance, is seen by some as a period when much less scientific and cultural progress was achieved. These so-called Middle Ages are often associated with monks – the most highly educated people of their time – sitting in cold, dark monasteries, creating beautiful 'illuminated' manuscripts on vellum sheets. There are perhaps many reasons for a fall-off in the number of new breakthroughs during that period, but one simple explanation could be that the flow of information was significantly reduced due to a lack of paper.

The Romans used papyrus, which was originally developed by the ancient Egyptians. They spread the use of papyrus across western Europe, but with the fall of their empire the supply of papyrus dried up. The even older vellum (or parchment), which is made from the skins of animals, was scarce and expensive; it was reserved for only the most important documents of the time, such as grand religious texts and important legal documents. When the easy transmission of information that papyrus had allowed during the Roman period was halted, progress perhaps slowed as a result. It was not until around 1380 that a new flow of information started, driven by the arrival in Europe of wood-pulp-based paper, quickly followed by the movable-type printing press. This, in turn, helped to drive the Renaissance, which was followed by the introduction of the 'scientific method', the Age of Enlightenment, the birth of capitalism (first developed by the Dutch, then taken to Britain at the time of the so-called 'Glorious Revolution' of 1688), and finally the Industrial Revolution.

*

It was perhaps English inventor Thomas Newcomen's 'atmospheric engine' of 1712 – the first practical fuel-burning steam engine – that ushered in the Industrial Revolution. Growth in global economic output experienced a step change in the 1880s, with the expansion of the railways, the first electric-power generation plants, the broader distribution of daily newspapers (enabled by faster transport), and the spread of the telegraph. Later, with the much broader roll-out of electricity, came early signs of an information revolution.

However, even up until 1993 the storage and transfer of information by humans was limited. Estimates suggest that all the books, music, photographs, cinema film and video tape that existed at this time represents a total of less than 50 billion gigabytes of stored human information. That is a lot, but it is nothing compared to what was around the corner.

Six breakthroughs changed this:

1. **Information theory.** This showed how information could be broken down into simple ones and zeros, and then communicated efficiently, giving us the mathematical framework for high-speed digital communications. We will look at this further in the next chapter.

2. **Integrated circuits.** These provided the underlying technology for building powerful computers, solid-state storage systems and communications equipment, which could all be scaled up to keep pace with the coming explosion of information-sharing.

3. **A combination of lasers and fibre optics.** These allowed information to be passed at the speed of light over long distances. The introduction of fibre optic communications in the early 1990s ensured that large numbers of independent channels of communication could all share a single cable running under the oceans and provided the platform for allowing huge amounts of information to travel around the globe.

4. **The development of a standard 'protocol'.** This global language for how different pieces of electronic equipment, from different companies and from different countries, could all talk together had been developed for the ARPANET computer networking system, which was used to connect universities and research centres, initially in the USA, then later in Europe and eventually around the world. The protocol that computers use is called TCP/IP, which stands for transmission control protocol and internet protocol, hence the word 'internet'.

5. **The World Wide Web.** A breakthrough made by Tim Berners-Lee, who was an independent consultant working at the CERN laboratory in Europe, the largest particle physics research centre in the world. TimBL, as he is known, developed a software environment during 1989 and 1990 that allowed information to be easily, and visually, shared over the internet (with websites preceded

by the letters www). Describing this development, Berners-Lee said:

> Creating the web was really an act of desperation, because the situation without it was very difficult . . . Most of the technology involved in the web, like the hypertext, like the Internet, multifont text objects, had all been designed already. I just had to put them together. It was a step of generalising, going to a higher level of abstraction, thinking about all the documentation systems out there as being possibly part of a larger imaginary documentation system.[14]

This insight has allowed us to easily share huge amounts of information around the globe, using the internet. It is important to remember that 'the internet' is the communications protocol that allows machines to talk to each other, but it is Tim Berners-Lee's 'Web' innovation that lets humans use this system in such an intuitive and human-like way.

6. **The opening up of 'the Web'** so that anyone could use it. In September 1993, the Clinton–Gore US administration released a report proposing the creation of a 'nationwide information super-highway'[15] that would be built on top of the ARPANET and would be open to commercial companies and individuals. From 1993 onwards, commercial activity started to grow on the internet using Web browsers that were based on Berners-Lee's work. I remember getting hold of one of the very first Web browsers and connecting to the Web in the summer of 1994 – compared to today, there was only a tiny amount of information that you could find at that point.

Since 1993 the rate of growth in information travelling over the internet has been stunning. Much more information is now generated and shared every year than even existed before 1993. According to a report produced by US networking company Cisco,[16] global internet traffic totalled around 1 megabyte per second in 1992, just before it was opened for commercial activity, but by 2016 this had grown to more than 26,000 gigabytes per second (a 26-million-fold increase). This rate of growth suggests a doubling in internet traffic every thirteen months. Around 100,000 gigabytes per second of data was transferred over the internet in 2021, which is equivalent to approximately 3,000 billion gigabytes per year, or 3 zettabytes. Compare this to the approximate 50-billion-gigabyte sum total of all human information that perhaps existed as recently as 1993 and you can see we are now sharing about 150 times more digital information each year than even existed prior to 1993.

*

Artificial intelligence needs three ingredients: powerful computer hardware for processing; humans to provide the computer with a method for learning; and digital information to learn from. The Web gave us digital information. If you were to plot human information-sharing on a chart for the last 44,000 years, it would basically show a flat line at or just above zero for the whole period, which suddenly, right at the end, goes up almost vertically, from 1993, when the public Web was created.

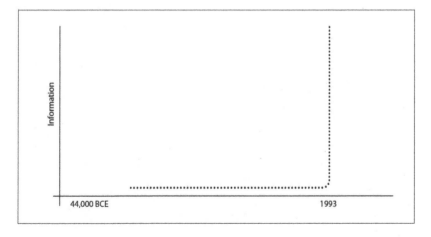

The accumulation of human knowledge

We are now very clearly living in the Information Age, and it is time to meet another of my technology heroes: the father of our Information Age, and someone who deserves much broader recognition.

7

CLAUDE SHANNON, THE FATHER OF OUR INFORMATION AGE

'Off and on I have been working on an analysis of some of the fundamental properties of general systems for the transmission of intelligence . . .'

CLAUDE SHANNON, IN A LETTER TO MIT PROFESSOR
VANNEVAR BUSH, FEBRUARY 1939[1]

Computers don't have fingers, so how do they count? You (most likely) have five fingers on each hand, and if you were an Ancient Roman you would count up to 'V' on one and then up to 'X' on the other. For some reason, 'X' times 'V' equals 'L', while 'X' times 'X' equals 'C', and so it is no wonder that as soon as we discovered the Hindu-Arabic numbering system, we all decided that this was much easier.

The decimal numbers that we use today have become standardized as the way to count. Like words, numbers are a way to represent pieces of information for communication. This use of numbers and fingers for counting has been going on for centuries. In the manuscripts (written on vellum) of the monk-scholar the Venerable Bede, which date from around 700 CE, we find various diagrams that depict counting and calculating techniques that use fingers, together with tables of calculations.

As I discovered in my first encounter with a computer, back when I was just eight years old, instead of fingers computers use binary digits or 'bits', a term coined by Claude Shannon, the father of our Information Age. A bit is something that can only ever exist as a one or a zero, like a switch that is only ever open or closed. Shannon wrote that any type of information can be represented as a set of bits. Today we take for granted that our music, our photos, and even documents and books, are now stored in this way. Bits are a very powerful concept: if you were to count using the binary number system that comes from these bits, using each of your ten fingers to represent one bit, you could count all the way up to 1,023.[2] Bits allowed us to build the digital world that we now inhabit, and it was the breakthroughs of Claude Shannon that allowed this all to happen.

In 1937, at just twenty-one years old, Shannon wrote one of electronic engineering's most important papers for his master's degree, showing how combinations of switches could be used to construct any logical or numerical relationship.[3] This paper provided the foundation for digital circuit design theory, which has subsequently been used to build all digital electronics, all computers and all semiconductors.

Feeding off this work, for my first electronics project as a child I built a digital clock. I used a quartz crystal to generate a reliable timing frequency and then, using Shannon's digital circuit theory, built logic circuits that divided this signal down to get seconds, minutes and hours. I also added an alarm function so that the clock could wake me up for school, but to make the logic design simpler I used switches that forced me to enter a binary number for setting the hours and the minutes of the alarm. No one else was ever able to set the alarm.

For Claude Shannon, his logic design theory was just the start. After completing his PhD in 1940, he spent time at the Institute for Advanced Study in Princeton, New Jersey, working alongside people such as the mathematician and logician Kurt Gödel and Albert Einstein, and then went to Bell Labs, the leading telecoms research centre, based in Murray Hill, New Jersey, to work on control systems and cryptography. All this time, Shannon had been working alone on a completely revolutionary theory: in 1948 he published his seminal work 'A mathematical theory of communication' in the *Bell System Technical Journal*.[4] The importance of this paper for communication systems and for our modern Information Age cannot be overstated.

As well as introducing the concept that all information can be described using bits, Shannon described a way to transfer information across distances digitally. Communication is the ability to reproduce information, accurately, at a new place. First you must encode the information into a message, then transmit the message over a channel, then receive the message and finally decode it so that the information can be faithfully reproduced. Shannon proved mathematically that information can always be reproduced at another point, even if there is a lot of noise in the channel.

Imagine you want to communicate with a friend. Normally you would speak to them and, since they speak your language, they would decode the information easily. If the channel for communication is noisy (perhaps you are in a very noisy room), you might choose to write the message down on a piece of paper and pass a note. If your friend is far away, you might put the note in an envelope and post it to them. In each of these cases you have transmitted your information using a different method of encoding (writing) and a different communication channel (a note that you hand over or post). These different encodings and channels might mean it takes longer for the transmission to occur, but the information has still been communicated.

During the communications process, the message transmitter and the receiver are trying to ensure that they can reliably share the same information. This idea of two entities sharing information is called 'relative information'. For communication to happen, these two entities must end up with the same information. Once the information that both parties hold becomes highly 'correlated' (i.e. becomes closely matched), it will have been communicated accurately. The purpose of communication is to reduce uncertainty (or entropy) until both the transmitter and receiver have the same information.

Sharing information allows us to gather knowledge and to share intelligence. As an example, the cells in our body share information so that they can work together to make our bodies function; even plants share information. We can also observe that the process of evolution is a communication of information from one generation to the next. Shannon's work also described the maximum theoretical speed at which information can be reliably sent over any given communication channel, a property that is known as 'the capacity of the information channel', or Shannon's limit. This is like a speed limit for information exchange.

Interestingly, I came across a great description of Shannon's ideas not in a book on information theory, but in a book on neuroscience: *Principles of Neural Design* by Peter Sterling and Simon Laughlin.[5] One simple example in that book shows the way Shannon's limit also applies in the evolution of a human brain. Let's imagine that your brain needs to send a single piece of information from one neuron to another. This piece of information will need to be converted into a message that can be sent to the other neuron (your brain uses a spike of low-frequency electrical charge). To send this piece of information, we need some way to pass the message between the two neurons along a link called the synapse, which is like a small 'wire' that forms the connections between your neurons. The purpose of the message is to reduce the uncertainty at the receiving neuron about what information has been sent. If the message signal strength is too weak, then the receiving neuron may not be able to 'hear' the bit of information correctly. If the message is sent over a long distance, or if the channel over which the message passes is too noisy, the message may become confused or garbled and this will mean that it is less certain that the information will be received correctly. If the message is sent too quickly, the receiving neuron may miss the message completely.

Your brain has evolved to be optimized for the accurate transfer of information between all the different neurons. The brain tries to keep close together those neurons that communicate most often. It limits the speed of messages in the brain to ensure that the information is received accurately. If the neurons are far apart, it will reduce the message speed or it will use more energy. Your brain also uses different techniques for sending messages. As well as the thin synapse 'wires' that pass the message as a low-frequency electrical spike, in some cases, for important information that must be shared across a large number of neurons, it will use wireless communication in the form of chemicals (such as adrenaline, which warns you of danger).

It turns out that our brains, and the brains of other animals, also conform to Shannon's limit. This speed limit explains why the amateur and professional tennis players have roughly the same reaction times: sending signals around the body is determined by Shannon's limit, and this physical limit is basically the same from one person to the next. When you look at the neural maps of insect and animal brains, they are optimized for the size and power available in these different-sized brains. An insect's brain is small, and the transmission of information is very efficient, requiring only a tiny amount of energy. A mammalian or human brain is much larger and less efficient, and therefore needs much more energy. In our case, a larger brain helps us survive, but the trade-off is that we need to be supplied with more energy in the form of food. The 'cooking hypothesis' first proposed by Richard Wrangham suggests it was our early ancestors learning how to control fire, and then using this to cook food, that allowed us to evolve our larger brains. He says: 'I believe the transformative moment that gave rise to the genus Homo, one of the great transitions in the history of life, stemmed from the control of fire and the advent of cooked meals.'[6] Cooking food greatly increases the energy yield and the speed at which you can consume calories and protein. Wrangham suggests that this led to the number of neurons in the brain roughly doubling between Homo erectus and Homo sapiens. Yet, despite having more neurons, our much-improved Homo sapiens brain still has a speed limit.

Let's look at a machine example. Imagine that you are trying to receive some information over Wi-Fi on your laptop. If you are a long way from the Wi-Fi router, then you may notice that the communication speed is lower. If you are in a different room, especially if your house has thick stone walls, you might notice that the signal is weaker, and this reduces the communication speed. If you live in a city and all your neighbours have Wi-Fi, then your Wi-Fi signal might suffer from interference, and this will also slow your Wi-Fi's speed. Claude Shannon anticipated all of this, decades before we had Wi-Fi. The fact that our Wi-Fi systems work at all in these situations is because of Claude Shannon. Companies use the maths that Shannon developed to design better communications technology that now carries information around our house and around the world.

A more subtle part of Shannon's insight was that he realized most human information contains a lot of redundant data. As a child, he built a telegraph system using the barbed wire that ran between his friend's house and his own. He knew that letters could be represented as a small number of dots and dashes using Morse code. To reduce the number of clicks on the telegraph device, Samuel Morse had cleverly chosen the most frequently used letters to be represented by the fewest dots or dashes. What Shannon understood from this is that you only need a small subset of bits to recreate the original information. Your digital music is often compressed as an MP3 file, and this encodes the digital music to dramatically reduce the number of bits needed to capture the musical information. This in turn makes the music easier to store and faster to stream. Using fewer bits allows you to transfer more information, faster. Shannon provided mathematical proof to show the minimum number of bits that are required to accurately recreate the original information in different communication channels.

Importantly, Shannon also recognized that you could use codes (a form of encryption) to protect the information from interference. The words that we use in language are a form of encoding that helps to improve the speed and accuracy of our communication. As an example, scientists and lawyers use lots of specialist words (a more complex form of encoding) to help them quickly share complex information.

Building on this idea, Shannon predicted that by using maths you could create perfect codes that would ensure all the underlying information gets through. The combination of turning information into bits, compressing them, encoding them, transmitting them, then decoding them accurately at the receiving end, together with a method for checking the result, enables you to reliably communicate information at the maximum speed, allowing communication systems to approach the theoretical maximum – Shannon's limit.

When I was young, I found it difficult learning to read. It turned out that I have a mild form of dyslexia, a common neurobiological variation that one in five people experience to varying degrees. People with dyslexia find reading and spelling complex words like 'dyslexia' difficult (the irony was not lost on me). Most people use the left hemisphere of the brain to decipher letters and words, which is also the part responsible for language. Dyslexics, however, use the right hemisphere of the brain, the part that is responsible for spatial understanding. With training and time, dyslexics can subtly rewire their brain to compensate. This neurodiversity can bring its benefits. As an example, I have always been fascinated by numbers and equations, and can look at a page full of them and almost instantly see any that are wrong. As my friends and family also joke, I am never lost: I appear to have a very strong spatial awareness, perhaps because the right side of my brain is working harder than the left. When, as a child who was still struggling to read, I discovered that 'clever' computers also can't read – they use a special form of numbering system – I found this fascinating. I have subsequently discovered that neurodiversity is a great strength, and that rather than looking at the world through other people's eyes, we should all try to see the world through other people's brains.

Dyslexia is genetically inherited. For his doctoral dissertation of 1940, titled 'An Algebra for Theoretical Genetics',[7] Claude Shannon spent time investigating this subject. His mathematical theory of genetics was written before the details of DNA were discovered and his work was not widely published until 1993, but it anticipated aspects of genetics that subsequently took years to discover. He used mathematics to study how different gene combinations will pass through several generations in a family. His work on genetics clearly showed how the communication of information underpins all of evolution.

A true polymath, Claude Shannon is remembered for riding around on his unicycle while juggling, and for the way he would think through a problem. In describing how his brain worked, Shannon said: 'I just see patterns.'[8]

*

We have now seen how the AI revolution became possible. The performance of electronic computers was turbocharged by the rapid improvements in semiconductors over the last sixty years. Software has advanced and now allows us to build complex applications that are run on these powerful silicon machines. The internet has made huge amounts of digital information available. It is this combination of powerful computers, software and information that has allowed AI to develop. By focussing on 'induction'-based approaches that use artificial neural networks, we have now started to build incredible AI systems. To learn more about how AI thinks, we should now look at how AI is different from our own human intelligence.

HOW IS AI DIFFERENT FROM HUMAN INTELLIGENCE?

I have always been convinced that the only way to get artificial intelligence to work is to do the computation in a way similar to the human brain. That is the goal I have been pursuing. We are making progress, though we still have lots to learn about how the brain actually works.[1]

GEOFFREY HINTON, AI RESEARCHER AND
TURING AWARD WINNER

8

WHAT IS INTELLIGENCE?

In May each year our family always looks forward to the arrival of the house martins. Someone will usually spot an advanced party that comes first to check if everything is OK, and to see what has changed since last year. The main group then follows, flying in from North Africa to our house in the English countryside. We always enjoy hosting these small birds, who build and maintain their nests under the eaves of our house and search for food in the fields around us while brooding and raising their two sets of chicks. When out walking the dog, I will see them swooping over the fields in the summer evening sunshine, catching insects in their beaks and chattering away. Flying lessons are always exciting to watch, though sadly every year one or two don't make it. Then, after a summer of them sharing our house, with us watching them raise and train their young, we see them congregating on the telephone wires, discussing their plans, and perhaps deciding on the best route back to North Africa. One day towards the middle or end of September, we wake up and they are gone.

There is no doubt in my mind that house martins are intelligent creatures. They have physical abilities that we humans find impossible – flying being just one. They also build nests, and reuse them year after year, with a DIY ability that far exceeds my own. They appear to be acting and communicating as a group. They navigate over long distances and come together as a flock from different wintering locations, returning each summer to raise their young in the same spot where they themselves were born. I don't need to anthropomorphize their abilities in any way; these are intelligent, self-sufficient animals that are highly evolved. Perhaps the house martins look down on me as they fly around, judging my inability to fly and catch insects in my mouth as a lack of what they might call intelligence. The way we describe intelligence is a human construct that is very much defined by our own human abilities. Perhaps house martins have a different definition.

As humans, we are very focussed on intelligence because we feel that it sets us apart from other species. We even called ourselves Homo sapiens – wise humans. We don't have the largest brains – elephant and blue-whale brains are much larger – but we do appear to have the most neurons.[1] We view our intelligence as superior, and as a result we want to measure it so that we can prove to ourselves how clever we are. We find many ways to compare our intelligence – in exams, through IQ tests, by earning academic titles and through institute memberships and awards. We tend to respect people who appear to display intellectual ability, especially in subjects that we find hard ourselves.

As a society, we associate intelligence with things that educated humans find difficult: playing complex games such as chess or Go; doing calculus; or providing critical analysis of a Shakespearean sonnet. These are obvious applications of conscious reasoning, but thought takes many subtler forms, such as interpreting sensory input and guiding physical actions.

It is perhaps starting to become clear that computing machines work differently from our brains. We can easily see patterns, but computers can't. A computer needs to be told what to do, step by step, in a program. It can't do anything at all before a person, or a group of people, create a program that tells the computer what to do. But this is where artificial intelligence is a bit different.

A computer program describes a logical method that, step by step, tries to solve a problem. As the problems become more difficult, even with much more computing power and much more information, it becomes harder and harder to describe the problem as a logical method. Writing a *logical* computer program that could win at Go is just too difficult.

However, we can describe a *learning* method that shows the machine how to learn from information. Machine learning, which underpins AI, is the process through which we describe a method for how a machine can learn from information to find a solution to a problem. Correctly constructed, a machine learning method can start to use the information we provide to build knowledge that can in turn solve the problem. At a very simple level, AI helps machines to recognize patterns, very complex patterns, and allows a computing machine to work much more like our own brains.

As we've seen, to make a machine learning method work, we need lots of information and lots of computing power. The World Wide Web has given us large amounts of digitized information, and major advances in semiconductors have given us the level of computing power that is needed. Now humans can develop AI solutions that solve problems using a computing approach that is analogous to human neural networks. We can now solve problems that were previously out of reach.

The underlying idea for how computation in the human brain might work was first proposed in a 1943 paper by mathematician Walter Pitts and neurophysiologist Warren McCulloch called 'A logical calculus of the ideas imminent in nervous activity',[2] which built on the early theoretical work of Alan Turing. The authors showed that simple elements connected in a neural network, like the neural network in our brain, can provide massive computing power. This same neural-network approach is now being used in artificial intelligence.

Not only is AI able to recognize objects in pictures, beat humans at difficult games and understand language, but it is helping us to solve some of the hardest problems in healthcare and allowing us to make new breakthroughs in science, and may also help us deal with climate issues. Your electronic calculator is a tool that helps you solve arithmetic problems. Your laptop is a tool that lets you solve accounting problems using a spreadsheet or write a book using a word processor. But AI is a tool that can help us solve higher-order problems that we previously found intractable.

Sometimes the problems that you are trying to solve have no simple, logical answer – they may be questions to which it would be meaningless to respond 'yes' or 'no', or to offer a number solution. Instead, for many complex problems that we solve every day, the answer ends up being a well-informed, highly accurate, probabilistic judgement – an impressive display of intelligence that we commonly call a guess.

We do this all the time. As in the case of our tennis player, there is insufficient information available to provide a completely accurate answer for how to strike the ball, but a good estimation of its speed and direction allows them to hit a return shot that is accurate enough to win the point.

This 'probabilistic' nature of humans and artificial intelligence also means that they will both sometimes get the answer wrong. 'To err is human . . .' the proverb goes, but with AI we will perhaps need to modify this to become, 'To err is a result of probabilistic judgements made with insufficient information by both humans and machines.' Perhaps not as snappy – but more correct.

This means that, unlike our electronic calculator, AI will sometimes get the answer wrong. It will 'err', but if designed well, it should 'err' much less than a human. Nevertheless, it will occasionally 'err', and we must be aware of this if we intend to use AI as a tool for good. It is worth keeping in mind the prescient quote from Alan Turing that I shared in the introduction: 'If a machine is expected to be infallible, it cannot also be intelligent.'

The potential for AI to help us solve very complex and very important problems is enormous. Having machines learn from information – rather than just telling them what to do, step by step, in a program – unleashes perhaps the most important breakthrough in computing since the very first electronic computers. However, we must also be cautious and recognize that, like humans, AI is not perfect; it may occasionally make mistakes. Also like humans, it may sometimes be led astray by the information that it is learning from; bias is possible.

AI has the potential to drive incredibly positive outcomes, but in the wrong hands could be used for harm and so we must take care. For one thing, we should try to learn more about our own human intelligence. To do this, we must first find a better definition of intelligence.

9

MORE INTELLIGENCE

'It is customary to offer a grain of comfort, in the form of a statement that some peculiarly human characteristic could never be imitated by a machine. I cannot offer any such comfort, for I believe that no such bounds can be set.'[1]

ALAN TURING, 1951

Albert Einstein said, 'The formulation of a problem is often more essential than the solution,'[2] and two examples from the development of technology show the truth in this. One clear example of a formulation of a problem is the semiconductor pioneer Gordon Moore's prediction that 'integrated circuit technology will double in capability every two years'.[3] What we now call Moore's law has driven the semiconductor industry to deliver a 25-billion-fold improvement over the last sixty years in the semiconductor devices that now power your computer and your mobile phone. However, Moore's law is not a law of physics. It was both a prediction but also a challenge laid down for the semiconductor industry to double the number of transistors every few years. It described a problem that Gordon Moore believed the industry could solve.

In the same way, a mathematical theory that Alan Turing captured in his 1936 paper 'On computable numbers . . .',[4] in which he described his theory for a 'universal machine' (what we now call a Turing machine), essentially described a problem that, if solved, would deliver a general-purpose computing machine. This incredibly important theory was the formulation of a problem that has subsequently allowed us, through experimentation and good engineering, to build the computers that we now use every day.

If you can define a problem clearly 'its solution may merely be a matter of mathematics and experimental skill', as Einstein concluded.[5]

The challenge we have is that intelligence is a very poorly defined human construct. Everyone seems to have a different description of what intelligence is.[6] Many of the descriptions have some anthropomorphic association, and there is no generally agreed-upon robust definition.

To help formulate the problem, as Einstein would have put it, let me suggest a possible definition of intelligence. My definition is deliberately very broad. However, it has the advantage of being a definition that can be applied to house martins and to all biological life, even to plants – and is not limited to humans:

> Intelligence is the ability to gather and use information, in order to adapt and survive.

How well species learn to adapt in different environments, and how well they survive and prosper, defines their level of intelligence. Of course, an artificial intelligence doesn't need to survive; it is just a tool, made by humans. AI is here to augment our own human intelligence. It gathers and uses information, and even adapts. But there is one crucial difference between artificial intelligence and the broader definition of biological intelligence: an AI's 'objectives' are defined by us. It serves our purpose, helping *us* to adapt and survive.

Within this broad definition of biological intelligence, humans, animals and plants must also have numerous second- and third-order capabilities, such as the ability to:

- sense and gather information to understand their environment
- learn how to acquire energy through food, nutrients, sunlight or

from other sources (including but not limited to harnessing the wind, harnessing the power of water, burning hydrocarbons, those captured from a chemical process or a nuclear process)

- generalize from information
- make predictions based on information
- recognize patterns in information
- identify associations between different pieces of information
- use information to solve problems
- communicate information to pass on knowledge and intelligence
- undertake planning
- generate innovative ideas and new information through creativity
- take appropriate actions (which may include motion to find food or to avoid immediate danger, or actions to deal with pathogens and disease, etc.)
- acquire knowledge through learning from information and through learning from experience
- store and recall both information and knowledge
- place the appropriate level of attention on key information
- show self-awareness by being able to reason and order one's own thoughts
- and perhaps most controversial and difficult to define of all: to exhibit consciousness by being present in the moment.

Homo sapiens would score well on this measure of intelligence. We make tools, construct buildings, and craft clothes that help us adapt to different environments. We grow crops and have domesticated animals for food. By capturing energy from different sources, we have built an advanced society. Also, we have developed advanced technology, such as the internet, that helps us cooperate and share information across the whole planet.

This broad description of intelligence also has the advantage of recognizing other abilities, such as sporting prowess and practical skills, and includes these as signs of intelligence. Our proposed definition for intelligence is not narrowly restricted just to those intellectual activities that only a few experts are good at.

It also brings into sharp focus human weaknesses – for example, our continuing disregard for the planet and its natural environment. As James Lovelock, one of the early thinkers in climate change, said in 1972, 'all living organisms form part of a self-regulating system that originally created and now maintains our environment and makes life on Earth possible'.[7] I am not sure that I would go this far, but the warning lights from our environment are clearly flashing bright red. Continuing to ignore these signals demonstrates an obvious lack of intelligence.

The issues that come from our propensity to seek dominance over each other also start to look less intelligent as a result of this definition. We have built weapons that, if used, would assure our mutual self-destruction and show humans' tendency towards hubris. We think of hubris as dangerous overconfidence, and it's often synonymous with arrogance. The ancient meaning was 'a transgression against the gods', and hubris was thought to be committed when a mortal claimed to be better than a god in a particular skill or attribute. The line between intelligence – where we use information to adapt and survive – and hubris – where we overreach – is often visible only with hindsight. As we consider how technology, in the form of artificial intelligence, can be used to enhance our own human intelligence, we must be careful not to overreach. Instead, perhaps we can use AI to help highlight when we are in danger of hubris.

A key component in this proposed definition of intelligence is the ability to understand our world, and here humans have some major advantages when compared to a machine.

UNDERSTANDING OUR WORLD

When you eat an apple, you reach out to touch it, noticing the colours, its shape and its waxy skin. You pick it up and feel its mass. The way it interacts with gravity allows you to judge its weight. You bite into it and hear the crunch as your teeth puncture the skin. You taste the sweetness in the flesh, sense the texture, and smell the fructose that has been built up by the sun's rays. Your digestive system identifies its energy-giving potential and signals to your brain and your emotions that this fruit is good.

Your brain is receiving information from all of your different senses and combining these inputs to build up a comprehensive understanding of this apple. You will also be learning about the environment in which you found this apple and about where you are eating it. Since you're enjoying the apple, you will memorize this experience so that you can repeat it. Inevitably you will compare this apple with others that you have eaten. This will let you build a more general understanding of all apples. You can then decide if this particular one is your favourite.

Multisensory learning is not unique to you. A dog does the same, as do cats and house martins. The learning methods of creatures are far more sophisticated than the simple, single- or dual-sensory training methods that have so far been used to create artificial intelligence machine learning systems.

Even when major pieces of information are missing, you can still use your prior knowledge of the world to build a much more complete understanding. You might even create a story that adds to what you are sensing. The famous French photographer Henri Cartier-Bresson felt that a photograph can capture this in what he called a 'decisive moment'.[8]

A two-dimensional black and white photograph contains much less information than our human binocular vision can capture from our constantly changing, three-dimensional, colourful world. However, shadows, reflections and areas where the image is in focus or out of focus can add depth and interest. The leading lines (for example, the masonry design and the way the light falls in the picture opposite) attract the eye and cause you to scan the image (notice how your

AI does not see in the way that humans do

eye is attracted towards the person even though they are not in the centre of the frame). The position of the photographer can also add drama to the photograph (for example, in this picture you are seeing a reflection, making the person look like they are perhaps part of a shop window display). Motion could also be captured by blurring in the scene.

The photograph above also asks questions that suggest a story. Why is the person standing there? Perhaps they are waiting for their wife who is looking inside the shop? He has a walking stick and appears to be leaning against the wall – is it difficult for him to stand like that? Our emotions and our understanding of the world combine with the information in the photo to expand our understanding and to create an emotional reaction. By contrast, an AI system might recognize 'person with walking stick'.

The human brain is obviously using information that comes from each of the senses. We then combine this information with our stored knowledge of the world. We add context that comes directly from the information we have just received, but also that comes from our broader understanding of the world around us plus knowledge from our memories. The brain is also feeding in emotions and feelings that direct our thought processes and attention. Humans appear not just to mechanically process received information but also to use new experiences to spark stories and memories that allow the very specific information received to be generalized and combined with our prior experiences.

But what happens when we receive information that doesn't match with our experience? In December 1976 psychologists Harry McGurk and John MacDonald wrote a paper in *Nature* called 'Hearing lips and seeing voices'.[9] By accident they had discovered a strange phenomenon while studying how infants perceive language at different developmental stages.

A technician accidentally dubbed the wrong audio on to a video. When this video was played back, instead of hearing the incorrect sound from the wrongly dubbed soundtrack, McGurk and MacDonald actually perceived a third sound that was completely different from both the correct sound and the incorrect one. Instead, what they interpreted was a sound their own brains had invented. They found that if the phonemes 'ba-ba' are played over the lip movements of a baby saying 'ga-ga', then what your brain might perceive instead is 'da-da'. What you are seeing and hearing doesn't align with your experience, so your brain will become confused. The third sound, the one these researchers actually perceived, was the brain's attempt to resolve this cognitive dissonance. Our multisensory system is both subtle and powerful but can also become confused. In these cases, our brain weighs the new evidence coming in against our accumulated knowledge and experience, and judges what is most likely.

In studying multisensory perception, researchers have studied different families of butterflies and tested their response to the size, colour and smell of various flowers.[10] Butterfly species respond very differently to these visual and olfactory (smell) cues. Four different responses were discovered from seven different species. In some butterflies, vision took priority over smell; in other cases smell took priority over visual cues; in two species no difference could be found. Some of the butterfly species who were driven more by visual cues were also influenced by the size of the flowers. Big butterflies preferred big flowers – perhaps because they need more nectar to support their larger body size. Meanwhile, the *Danaus genutia*, or common tiger butterfly, was not sensitive to colour at all, just to smell. It may seem strange to us, but perhaps this butterfly is colour-blind. Next time you see a butterfly perhaps you will notice which senses it seems to be using.

EXPERIENCING CHANGE

Chew on a piece of gum and it will quickly seem to lose its flavour as your saliva dissolves the sugars (or sugar substitute) in the gum. The interesting effect here is that the chewing gum hasn't actually lost its taste; you've just become less interested in the flavour, and as a result less able to sense it. If you were to take the gum out and leave it for a short period, when you pop it back in your mouth you will notice the taste of gum again. As the initial impact of the sugars faded, your brain stopped paying attention to the taste and smell of the gum, since there was no new information in it.

Your brain is giving more attention to the rate of change in the taste of food. When the taste stops changing, or the impact fades, your brain reduces the amount of attention it gives to these inputs. These temporal effects are very important to the way we learn and understand the world around us. The combination of colour, smell, taste and texture are all things that top chefs play with to stimulate your senses when creating new culinary experiences. Eating is not just about taste; it is a multisensory experience that comes together in your brain. The award-winning chef Heston Blumenthal from the Fat Duck restaurant in Bray, UK, has a research lab where he and his team develop new culinary sensations that challenge the senses. Egg and bacon ice cream, anyone?

Now let's test your visual system. Stand up and survey a scene. Try to take everything in. You'll notice that your eyes are constantly scanning around, focussing on one thing and then another. Your eyes don't work like a smartphone, taking a single picture. Instead, you build up a whole patchwork of different images and stitch these together in your brain to understand what's in front of you. Your eyeballs are constantly moving to find new information, and they focus their attention on what has changed. This lets your brain build up the complete picture of the scene.

This focus of attention only on the elements that change has also inspired how digital media files can be stored more efficiently. A video compression system identifies what has stayed the same in the frame and what parts have changed. The compression system only stores the information that has changed from one frame to the next, which dramatically reduces the total amount of stored information. Then, to reconstruct the images on playback, it builds the background and adds the changing pixels from each frame. The result is roughly a ten- to fifty-fold reduction in file size.

The photoreceptors in our eyes and the neurons in our brain appear to operate in a way that is similar to these digital compression systems. We memorize images, and the objects in them, as highly abstracted concepts that all build to establish our view of the world. The benefit of this approach is that our brain dramatically reduces the amount of information required both for processing and for storage in our visual system.

COMMON SENSE

Perhaps you recently bought a new electronic device and had to figure out how it works, or you had to navigate your way through a foreign airport at a new destination. How is it that you knew how to act even in these new situations? Humans, and animals, appear to be able to learn about how the world works just by observing what is happening around them. Occasionally we will combine these observations with just a small amount of physical interaction to test our assumptions. We appear to have the ability to quickly update our model of the world. We generalize new information and apply it in new situations to work out what is happening and what might happen next.

We tend to describe this ability as common sense. All animals appear to predict consequences and use knowledge models that they have learnt from information. They use them to reason, to plan and to explore. You build world models and then use them to imagine solutions to these new problems. But perhaps most importantly, these models allow you to recognize danger. The alarm will go off when you come across a complex situation that you have never experienced before. Often, before you consciously pinpoint the contextual factors, you will be alert and on guard. Common sense helps us to survive.

Common sense lets us predict what might happen next, but we also use this skill to continuously fill in missing information. We use new information to help us expand our world models. We learn how to deal with a new environment by generalizing from a prior learnt model and then we apply this to a new situation. Just as we have a hierarchy of memory systems, we also appear to build a hierarchy of many different world models that are situationally specific. You will have an accurate model for a place you know well – finding your way around your bedroom in the dark, for example – but for other places, such as a hotel room you've never stayed in before, you will quickly build a more rough-and-ready model based on situations you have seen before.

This also leads to surprise – for example, if you return to an old, familiar part of town where you once lived and turn the corner to find that part of the street has been redeveloped. Surprise is a discontinuity in one of our world models. Humour plays on this, as do many of the best films and stories.

Starting at birth, and then over the first few days, weeks and months of life, babies start to build these models. They will learn with every touch, and from what they see, that objects sit at a distance. Very quickly they start to realize that they are in a three-dimensional world. Watch how a baby moves its head when looking at an object for the first time. You will see it looking from different angles. It is using its two eyes to try to understand distance and to build a better picture of the object.

Once babies start to know about objects, they quickly learn to collate them into broad groups. They realize that objects don't just suddenly appear, disappear or magically change shape. They see that some objects move, and some are static – a quick push will help in building their knowledge of the world. They will try to interact with objects and learn that some respond. Some interactions work better than others – the cat might run away, but a grandparent will take their offered hand. Babies quickly start to build social skills such as smiling at people due to the positive response that this produces. They will also start to explore language by using their sophisticated vocal muscles to make different noises. They are building up their own models of the world and starting to develop their common sense.

It turns out that you help a baby to build their self-esteem when you encourage them to smile. They will start to realize that their actions and feelings are important. They are learning that they have the ability to affect their environment. This interaction will also help them to develop certain parts of their brain, specifically the areas that deal with emotions and empathy – how my actions affect you, and vice versa.

Later, when they grow up and become a young adult, they might learn to drive a car. After just twenty hours of tuition, they will move from being consciously incompetent to being consciously competent, with enough skill to pass their test. After a few more months of driving, they will start to become unconsciously competent – but unfortunately this is also when overconfidence can start to creep in and accidents can happen.

Even new learner-drivers use their models of the world to exercise caution and to understand danger signals. They quickly learn to slow down for corners and anticipate dangers ahead. However, these skills don't come naturally to an AI-powered autonomous vehicle. Even after receiving huge amounts of information from human experts, after training on millions of different reinforcement learning exercises in a virtual environment, and after being hardwired with key behaviours, they still struggle. Based on official regulatory data filed in California by the autonomous car company Waymo, an Alphabet/Google subsidiary, their autonomous vehicles drove over 2.9 million miles during 2022 in the state of California alone, all with a person sitting at the wheel ready to take charge. They reported an average of just over 17,000 miles driven between incidents where the human had to take over.[11] Developing common sense is very natural for humans but perhaps one of the hardest challenges for artificial intelligence.

OUR COMPLEX WORLD

As you may have noticed, in this book I am always careful to distinguish between the words 'information' and 'data'. A set of measurements is data, but without any context, such as 'these measurements are for a box', and without knowing some point on the box that relates all these measurements together, the measurements will remain just unrelated pieces of useless data. The context of knowing that the data relates to a box, and how the pieces of data relate to each other, turns data into information.

By analysing information, we can find patterns that unlock meaning. These connections and relationships between information provide knowledge. Knowledge can provide insights that allow us to make intelligent decisions. To help unlock this knowledge and apply it in our everyday lives, we simplify the world around us so that we can make sense of it. The world models that we use every day are only very rough approximations.

Our world is much more complex than we realize. We live in a space-time continuum, with time (as it appears on a clock) being a human construct that we use to simplify our interaction with the world. The world time standard that many countries use is set from London, and is defined as the average solar time measured at the Greenwich Royal Observatory, using the exact moment that the sun, at its highest point, crosses the Greenwich Meridian. However, because of the Earth's uneven velocity on its orbit around the sun and its axial tilt, noon is very rarely at this exact moment. And time actually travels imperceptibly faster if you live at the top of a hill compared with living at sea level as a result of something called 'gravitational time dilation'. So even time is an approximation.

As you are perhaps also aware, the light from a far-off star could be tens of thousands of years old. When you watch 'live' television on a screen placed on the wall at the other side of the room, the image has taken perhaps one tenth of a second to reach your home from the television transmitter, via a satellite. At the quantum scale, things become even less clear, with quanta, superposition and entanglement being incredibly difficult concepts to understand.

Even understanding how communication happens is rather complicated. For example, when I sit at the desk in my study I can see an oak tree through the window. I see this tree because photons have travelled 150 million kilometres from the sun on a journey that takes over eight minutes, and they have entangled with the atoms and cells in the tree. Some are absorbed and others bounce off, travelling to the photo-receptors in my eyes. My brain uses this to reconstruct the colour, shape and location of the tree, turning this data into information. The very first time I looked out of the window I learnt that the tree was there. It has now become part of my memories. If it were completely dark, I could still go outside and through my sense of touch learn more about the tree. The tree exists through a communication of information that happens between the photons that travel from the sun, the tree, and me.

I have clear information that tells me (and others) that the tree exists outside my mind. We are not straying into solipsism here. However, I cannot be certain that the tree will always be there. At some point in the future the tree may be blown down or hit by lightning. On a foggy day the channel that brings information about the tree will be fuzzy and the state of the tree will be less clear. The past may be knowable, but the future is not.

The physicist Werner Heisenberg showed that the amount of information in the universe is finite.[12] Everything can be boiled down to tiny, discrete pieces called quanta and you cannot go any smaller. Assuming there was some initial state (say, the Big Bang), this means that everything in the past is theoretically knowable. However, the future is not. The future is probabilistic and uncertain. Although we can make guesses and predictions, we cannot know for sure what will happen next. The future remains foggy.

The detail involved in the complex processes that underpin our world are so complicated that they are very hard to apply in everyday life. We efficiently ignore most of this complexity and instead construct descriptions of the world that are much simpler. We do this by defining constraints for the world around us so that it becomes a less complex place. We try to make the world a bit more understandable – the human construct of 'time' being just one example. Most engineers still use the classical Newtonian laws of physics, which work perfectly well for building bridges and skyscrapers. Civil engineers don't need to worry about Einstein's relativity or strange quantum effects. Even with a specialist understanding of relativity and quantum, our human understanding of the true complexity of the universe is still rather sketchy. To cope, we create further simplifications and approximations. Our intelligence is limited by the information that we receive and by our ability to process this information. It is limited by how well we build knowledge models, and by the way we convert this knowledge into intelligence.

Just as a human does not possess a completely general intelligence, capable of knowing everything, so an artificial-intelligence-powered machine can never become 'generally' intelligent. It is more efficient to constrain the environment that we ask a machine to understand, and we shouldn't ask it to try to learn everything. When AI is being trained, we should focus it on a specific domain and on a limited set of objectives. This will also allow us to test that the system is performing well. For example, if we have an intelligent robot working in a factory, it only needs to learn about this specific factory environment. It may be possible for the machine to become 'generally' intelligent within this limited domain and to know a lot about the factory and the products that it is being asked to handle, but that is all it needs to know.

It may also be possible for AI to become generally intelligent over a much larger domain and become a subject-matter expert that can help us to be more productive or replace laborious and dull work. But this general intelligence will still have some limits. Since an AI system learns through a method that we define, it cannot suddenly become fully sentient. It may know a lot because it can have access to very large amounts of information, and perhaps it will be able to use this information to know much more than a human can. It may become a real expert in key subject areas, far exceeding the knowledge of a human expert. A machine can already beat a world champion at chess and Go, and we should expect to see a machine that knows more about legal precedents than a leading barrister, or more about aircraft maintenance procedures than the most experienced engineer. There are also limits on how far it will be useful to go. Do we really need AI systems like Marvin the Paranoid Android, a character from Douglas Adams's *Hitchhiker's Guide to the Galaxy*? A humanoid robot instilled with afflictions of severe depression and boredom might be a useful foil for humour in a book but is not 'generally' very useful. AI systems should be designed to help us become more productive, allowing us to focus more on our human skills of curiosity, creativity and critical thinking.

GETTING A GRIP ON UNCERTAINTY

As we come to realize that our world is a complex place, we need to learn more about this concept of uncertainty and randomness, or to give it its more correct term: entropy. Uncertainty can be described and even defined mathematically as the level of randomness, or the level of entropy. But how can we quantify randomness? A random state is just a state of chaos. A random event is something that we could never have predicted. A random number is one that cannot be calculated. By definition, randomness is unknowable. The theory of quantum mechanics describes that we are surrounded by a world of tiny particles that are all governed by randomness. This idea of a universe governed by randomness upset Albert Einstein so much that he said: 'I, at any rate, am convinced that [God] does not throw dice.'[13]

Claude Shannon was fascinated by randomness. He understood that in communicating information, we are trying to convert randomness into structure. An analogue television or a radio that is not tuned in will just generate noise – what you are seeing and hearing is just random bits. If we can find the correct frequency and tune into the transmission, then this noise will turn into structured sounds and pictures – the information being communicated will start to become clear.

Driven by this interest in randomness, Shannon named his family home, near Boston, USA, Entropy House. Entropy is just a mathematical name for randomness. Apocryphally, Shannon asked the famous physicist and polymath John von Neumann what he should call the communication uncertainty that was central to his 1948 communications paper, and von Neumann is said to have replied to the effect that 'you should call it entropy for two reasons: first because that is what the formula is in statistical mechanics but second and more importantly, as nobody knows what entropy is, whenever you use the term you will always be at an advantage.'[14] Claude Shannon described communication as an 'entropy of bits'. When entropy is high we just receive noise, but by finding ways to reduce entropy we start to receive information.

Physicists also use the term entropy to describe the state of matter, and it is importantly tied up with the Second Law of Thermodynamics. Energy cannot disappear, but it tends to disperse over time; physicists call this an increase in entropy. At maximum entropy, matter becomes inanimate – the atoms have equal energy. An animal that is alive and vigorous, with clear structure, has low entropy, but when it dies and becomes inanimate it is said to trend towards maximum entropy. Entropy is reduced by adding an external source of energy, which in turn supports the communication of information. Energy and information work together to reduce entropy.

All around us we can see this entropy being resolved by biological organisms to make the beautiful and intricate structures of flowers and plants, which are built by cells that receive their energy from the sun. The cells in these plants are using energy to communicate and cooperate with each other, and to create these structures. We see it in the amazing insects and animals that feed off, learn from and adapt to their environment. And we see it in human activity that creates structure by putting up buildings, growing gardens and building cities that provide comfortable and ordered environments. We also set rules and define legal systems that allow our modern societies to operate in an orderly fashion, all made possible by the addition of energy and the communication of information. 'Organisms organize,' says the science historian James Gleick in his book *The Information*.[15] 'We sort the mail, build sandcastles, solve jigsaw puzzles, separate wheat from chaff . . . compose sonnets and sonatas . . . and all this we do requires no great energy, as long as we can apply intelligence . . . We disturb the tendency toward equilibrium . . . It sometimes seems as if curbing entropy is our quixotic purpose in the universe.'

The famous Nobel Prize-winning physicist Erwin Schrödinger also highlighted this point, asking: 'What is the characteristic of life?' and answering that it is 'When it goes on "doing something", moving, exchanging material with its environment', as opposed to becoming 'an inanimate piece of matter'.[16] As Schrödinger goes on to say: 'How does a living organism avoid decay? The obvious answer is: by eating, drinking, breathing and assimilating.' As he points out, the technical term used for this process in plants and animals is a word derived from both Greek and Old English: metabolism, meaning exchange – the exchange of material.

I have proposed that 'Intelligence is the ability to gather and use information, in order to adapt and survive.' We have learnt that information lets us build knowledge and that this knowledge provides us with intelligence. By combining the observations of Claude Shannon and Erwin Schrödinger, we can see that by using energy that they find, biological organisms can communicate and gather information. This exchange is a fundamental part of life, and in turn helps the organism build structure so that it can find and capture the energy needed to survive. The intelligence that results allows all biological life, including us, to survive and prosper as species. Energy drives intelligence; intelligence allows us to then capture more energy.

This means that a biological organism's purpose is to use the energy that it captures to communicate and learn from information so that it can build intelligence, which in turn will allow it to capture more energy so that it can then build even more intelligence to find more energy . . . and so on, and so on. Evolution favours organisms that are better at capturing the type of intelligence that lets them deal well with their environment. Intelligence helps us build structure and organize in order to push back against the chaos that would otherwise engulf us. By just describing this observation of the link between energy and knowledge in a slightly different order, we get 'Intelligence is the ability to gather and use information, in order to adapt and survive.' This is perhaps our overarching biological purpose.

10

CONSCIOUSNESS

Consciousness remains a mystery, certainly to me, but also to neuroscience. It is everything that you experience: the taste of strawberries and cream, a theatre production that made a great impact on you, but also the jarring pain from a muscle pulled during your morning run. 'The state or fact of being mentally conscious or aware of something' is one of the ways that the *Oxford English Dictionary* defines it. But 'consciousness' is such a loaded term and, like intelligence, almost everyone you ask will have a different description.

Often, consciousness is considered to be reserved for humans and is discounted in plants and insects, and even in other animals. As humans, we focus more on our very deliberate thought processes, which we think of as conscious thoughts or conscious reasoning, and we tend to ignore the hidden intelligent processing that is required to hit a tennis ball. We describe our ability to tell stories, where we are sharing a conscious experience, as an example of conscious thought, but animals tell stories too – even bees.

When a worker bee returns to the hive, it will be directed by other bees to parts of the hive that need more nectar. Once the nectar has been safely stored, this returning bee will attract the attention of others and then perform a waggle dance.[1] This bee dance is a coded description of where a new source of nectar can be found. By dancing forwards and back, and then turning to the left or right, the bee's dance will show the route to this new source of food and energy. Bee dancing is far more than just a basic response. It includes the recall of information, marshalling of thoughts and the replaying of information. The pattern of the dance provides a clear communication to the other bees so that they can find their way to this new source of energy. Does this mean insects are capable of conscious thought and conscious reasoning?

In his 2011 book *Thinking, Fast and Slow*,[2] the Nobel Prize-winning psychologist and economist Daniel Kahneman describes how it appears that our brain operates using two systems, which he describes as the fast and responsive System 1, and the slower and more thoughtful System 2. We use System 1 to hit a tennis ball, and we use System 2 when playing chess or working out how to solve a cryptic crossword. We tend to overestimate the importance of System 2 and underappreciate the incredible ability of System 1. This is because, as Kahneman describes, when we get into difficulty with System 1, we tend to call on System 2 to help us out.

However, our System 2 brain is not perfect either, and relies on experience to achieve a good outcome. During the recent Covid-19 pandemic, patients were presenting at hospitals with breathing difficulties caused by the virus having entered their lungs. Doctors were unsure what the correct treatment was for this new virus. In Moldova and other countries, where tuberculosis still circulates among the population, doctors had seen similar lung-infection symptoms caused by these tubercular diseases. They would routinely prescribe steroids to help the body fight the lung infection, and so they tried the same with this new Covid-19 virus. The results were positive, and it was this broader understanding – built from experience – that helped them address this new problem.

A few years ago, our son became ill with a bad stomach ache. The doctor at our local surgery diagnosed appendicitis and so we took him to the hospital. Two junior doctors there confirmed the diagnosis, and he was scheduled for an operation. The procedure was delayed because an emergency case came in overnight. The next day, a senior consultant was doing the rounds and examined our son. This highly experienced doctor quickly decided that the condition wasn't appendicitis. He told the junior doctors to feel the stomach area and explained that it was not hard in the area he would expect for a swelling of the appendix. He was able to make a diagnosis using his prior experience. The other doctors had needed to extrapolate outside their area of direct knowledge, and as a result their conscious reasoning process was not as accurate.

Erwin Schrödinger touches on consciousness in the 1944 book *What is Life?*[3] which includes a set of papers called 'Mind and Matter'. He describes a particular contradiction that we can all perhaps relate to: 'biology shows us that our body functions as a mechanism controlled by the Laws of Nature, and yet from our own incontrovertible first-hand experience we all know that we are [consciously] directing our own individual motion. We can foresee the effect that this motion will have, and we take full responsibility for these actions'. Schrödinger points out that 'consciousness is associated with the *learning* processes of a living substance'; whereas once we have learnt how to do something, the act often becomes unconscious. So, is consciousness just the process of learning or assimilating new knowledge?

When you start a new job, the first few days always appear to go by very slowly. You are meeting lots of new people, taking in lots of new information, learning your way around the office and finding new places for lunch. In your second week you start to click into the routine and time goes by more quickly. After a few months, everything just feels normal. The familiarity that you have gained about your new surroundings has allowed you to switch from being very consciously focussed on gathering new information (to build knowledge) to a more routine state of unconscious actions that are using this newly learnt knowledge.

Consciousness is also often linked with religious connotations. In Buddhism, for example, consciousness refers to your level of self-awareness and there are nine levels of consciousness that you can achieve. The first five relate to being fully aware of your five senses: seeing, touching, tasting, hearing and smelling. The sixth level is understanding the combination of all this information that is coming from your senses. The seventh is recognizing the inner abstract thoughts, which are distinct from your senses. The eighth is linked to memory and is the accumulation of all your thoughts, words and actions. Buddhists believe that this 'storehouse' of knowledge and memories will persist after death and into reincarnation. The ninth and highest level is the purest level of consciousness and is the foundation of your life. By rising to this highest conscious level, Buddhists believe that you can clear yourself of negative karmic energy. They believe that negative energy will produce misery and lead to suffering, either for yourself or for others. By becoming fully conscious, and being fully aware of the moment, you can lead a more fulfilled and enlightened life.

A computational neuroscientist might look at things from a different viewpoint. David Marr first studied mathematics at Cambridge University, and in his 1969 PhD thesis he went on to explore the new subject area of how the brain performs computation. He captured detailed anatomical information of the cerebellum, the neocortex and the hippocampus, trying to understand how they work. Marr developed a deceptively simple three-layer computational framework for tackling these challenging neuroscientific questions.

The first and highest layer that Marr described is the 'computational level', which sets out the goals of the system and describes what outcome the system is trying to achieve. The middle layer is the 'algorithmic level', which defines the processes and algorithmic computations that will allow the biological system to achieve these goals; this layer describes a method, in the same way that AI describes a method for learning. The third and lowest layer is the 'implementation level', which describes the precise way in which these computations are performed. In a biological system, this layer is built from neurons and connections in our brain.

Marr's approach to dividing neuroscience problems into these different levels has become a cornerstone in the study of the brain. Marr tragically died from leukaemia when he was only thirty-five years old but his book *Vision*, published posthumously in 1982, went on to dramatically influence neuroscience and, more recently, AI research.[4]

Neuroscientists now approach problems either by trying to understand the computational level – the outcomes that are trying to be achieved – and then working down, or they start by studying the neurons and synaptic connections and then try to work up. Even among neuroscientists, though, there is no universally agreed description of how consciousness works in the brain. Accurately measuring neural activity is difficult and no one has yet built a consciousness meter – although some are trying.

One major theory of consciousness is called the 'global neuronal workspace' (GNW), proposed by psychologist Bernard Baars along with neuroscientists Stanislas Dehaene and Jean-Pierre Changeux.[5] They start from the implementation level, with the idea that the neurons' 'shared workspace' causes consciousness. As Baars explains, 'we capture information from a momentarily active, subjectively experienced [event], and then store this in a working memory'. He describes that 'in the inner domain of this working memory we can rehearse telephone numbers to ourselves or we carry on the narrative of our lives'. He says that this working memory 'includes inner speech and visual imagery'. This theory proposes that once we can understand the low-level implementation our brains use for this working memory, then computers in the future will also be able to implement the same type of low-level computations. As a result, Baars, Dehaene and Changeux believe that computers could become conscious.

Another group of neuroscientists, including Giulio Tononi from the University of Wisconsin, have a top-down theory. This starts with experience and then tries to work down to see which parts of the brain are activated by these experiences. Tononi calls this the 'integrated information theory' of consciousness (IIT).[6] According to this idea, each of our conscious experiences is unique and specific to us. For example, when I walk my dog through the field, as the morning sun is just coming up and with the grasses in the meadow blowing in the wind while a kestrel hovers, ready to dive and catch its breakfast, this unique experience cannot be broken down into its different parts without the overall experience ceasing to be what I actually felt. This particular experience is unique and is owned by me. Tononi and the proponents of IIT state that any complex and interconnected mechanism that can encode this set of cause-and-effect relationships will capture something. But if the mechanism is missing the integration and complexity of all the incredibly subtle parts that make up the total experience that I might be feeling at that specific moment, then it becomes just a description. The machine has not actually experienced it in the same way that I have; instead it is simply like a camera capturing a photograph.

Unless a machine can mimic all the different sensory inputs and emotions, mixed in with my previous memories and experiences, the machine will not be able to share the same conscious experience. As a parallel analogy, it is possible to simulate the effects of the enormous gravitational fields of a black hole on a computer. But this simulation does not actually cause any gravitational effects near the computer. In the same way, a computer may be able to simulate something that approximates to conscious thought, but it won't actually experience anything. IIT states that 'consciousness is an intrinsic causal power associated with a specific complex mechanism'. Human brains are unique, and although we might eventually be able to simulate some form of sensory experience in a machine, it won't be consciousness. The system will just be following a method that we have imprinted into the machine, and it will be describing an experience in the same way that a parrot might repeat words that we have taught it to say. Today, AI systems like ChatGPT act as stochastic parrots,[7] just repeating words and sentence structures that they have learnt without having a real understanding of their underlying meaning.

One thing that neuroscientists do agree on is that, intriguingly, there is an area in the brain that appears to be responsible for our conscious thoughts. We know that if the spinal cord or nervous system is damaged through a major trauma, even though this may result in paralysis, a rich variety of life can still be experienced and enjoyed – conscious thought is not affected. The cerebellum, which sits at the base of our brain, is considered the most ancient in evolutionary terms, and is the piece of our cerebral matter that has by far the most neurons. The massive number of neural pathways located here control our highly complex sensorimotor abilities, allowing us to balance and walk on two legs, and then run and strike a tennis ball. It is clearly linked to what Daniel Kahneman would call our System 1 brain. Neurosurgery in this region, such as for the removal of a tumour, may affect our balance or our ability to play the piano, but it appears to have no effect on our ability to sense and experience life.

Another major part of the brain is the highly wrinkled outer surface, which is called the cerebral cortex. This complex area of the brain holds about 15 to 20 per cent of our neurons. The many folds and valleys in this outer layer provide a larger surface area, with many more complex neural connections than in other parts of the brain. These connections allow much more information to be stored and quickly processed. This is the area that is associated with our higher-level processing, such as seeing, spatial awareness, hearing, language, thought, emotion, reasoning and memory. It is also the area that neuroscientists identify as vital to our consciousness.

Experiments by Tononi and his colleague Marcello Massimini have shown that conscious thought is associated with an area that straddles three of the four lobes that form our cortex. The areas identified comprise the parietal, temporal and occipital lobes. The main activity in conscious thought appears to be concentrated in an area they describe as the 'posterior hot zone'. By using a pulse of magnetic energy, they were able to induce a brief electric current in the neurons of the brain.

A set of electroencephalogram (EEG) sensors positioned around the skull were able to measure these electrical signals to build a three-dimensional picture of the activity. From this information they were able to estimate the complexity of the brain's response in these different regions and identified this posterior 'hot zone' as the source of conscious thought. When people were asleep, the complexity was low, and when awake and active, the complexity of response in this region was high. In tests on patients with brain injuries, but who were responsive and awake, the method was able to confirm evidence of consciousness in this region. Tests on forty-three brain-injured patients, where all attempts to establish communication had failed, confirmed that thirty-four were indeed unconscious, but nine were not, suggesting that very sadly these nine patients were aware and conscious but remained unable to communicate with their loved ones.[8]

Trying to understand consciousness is difficult, and it is perhaps impossible to break it down into a distinct set of computing steps. This is what the IIT (integrated information theory) in neuroscience also suggests. The philosopher David Chalmers, in his paper 'Facing up to the problem of consciousness',[9] proposes that consciousness can be divided into an 'easy problem' and a 'hard problem'. He describes how the 'easy problem' of consciousness can be explained in terms of computational or neural mechanisms, and includes:

- discriminating, categorizing and reacting to the environment
- integrating information from our different senses
- describing our mental state
- deciding what to focus attention on
- performing conscious control
- the difference between when we are asleep and awake

He describes these 'easy problems' as relating to actions or 'functions' that can be broken down into a simpler set of underlying processes through a reductive process. Using David Marr's approach, if we can explain how something works then we have also explained the higher-layer function. For example, if we can explain how our eyes build up a picture of the world around us then we have described how we see our environment, and we can then go further to identify the underlying neural processing that makes this possible. The same is true for the difference between when we are asleep and awake.

It is well known that there are different levels of sleep, and scientists have identified at least five states. These are described as: awake, N1, N2, N3 and rapid eye movement sleep (or REM). N1 to N3 are non-rapid eye movement (NREM) levels of sleep, with each representing a deeper state of rest. During some parts of sleep – especially during REM sleep – your brain is actually very active with dreams and other brain activity. As an example, when our dog sleeps he will sometimes make little noises and his feet will move like he is attempting to run – he is obviously having a wonderful dream. Depending upon your sleeping state, you will display different levels of consciousness. But even in the very deepest sleep state you can still be woken by a baby crying, by a touch or by a noise that might signal danger. You continue to have some level of awareness of your environment. This seems perfectly explainable – especially if you are familiar with the design of microprocessors.

The microprocessor in your mobile phone also has a very sophisticated set of sleep states, including something called a 'deep-sleep mode'. However, even in this lowest power mode, the microprocessor can still wake up within a few milliseconds when the correct signal is applied. The processor can operate in different levels of sleep, which will trade off on wake times and compute performance all the way up to the maximum performance level, when it will also be drawing the most power. If you regularly do video conference calls on your phone, for instance, you will discover that the battery will drain very fast. The combination of the screen being on, the cellular modem working at full speed and the processor handling the compression and decompression of the video information all results in a very large drain on the battery. Humans appear to have a similar power-mode system and have ways to reduce the power required in certain parts of the brain but with small areas remaining alert and ready to respond. You have perhaps experienced this yourself when walking with a friend if the conversation becomes intense: you will find yourself stopping so that your brain can apply more energy to this important discussion. Your brain will shut down the subconscious part that is controlling your muscles for walking.

The 'hard problem' in consciousness is to explain the subjective element of conscious experience, or why we react to and experience beautiful music, a lovely sunset or an amazing theatre performance. David Chalmers argues that our experience cannot be explained through a reductive process, by breaking these experiences down into simpler steps. He suggests that conscious experience cannot be explained other than as a new fundamental theory. He does, however, link experience to information theory and the work of Claude Shannon. Shannon described information states as being embedded in an information space.[10] This is a little hard to explain but, basically, an information space is a very complex high-dimensional environment, meaning that it is made up from many different variables, with each variable influencing the outcome. This information space helps us to compare new information with what we are already familiar with (in other words, finding complex associations between lots of different pieces of information). It also allows us to understand how different this new piece of information is compared with our previous experience.

As we saw in Chapter 9, in the way we experience chewing gum, changes in taste, texture and smell will provide us with a much richer experience. We also know that experience can be both positive and negative. When we hurt ourselves, the feeling we get is negative, but if we complete a difficult task such as a cryptic crossword puzzle, or find something aesthetically pleasing, then our experience is positive. Our multisensory mind is not only taking the basic information that we are collecting from our environment through our senses, but it is mixing this with our memories and with our emotions through an associative process. As we absorb this new information, we are considering the extent to which this information is new, how rapidly it is changing and how much it affects our emotions.

The 'hard problem' of conscious experience is perhaps a method that our brain is using to evaluate how significant any new piece of information is. As Schrödinger says, it is perhaps related to learning and we are deciding what to place attention on.[11] This will be influenced by how novel the information is, how rapidly it is changing and how positive or negative this new experience is. This is intimately linked to our emotions: our emotions help us understand how different or special this new information is. In any experience there will be lots of changing information, so working through all the variables will need a very complex set of circuits in the brain. Our brain will need to understand whether this information is filling a gap in our existing world view – known as interpolation – or whether this is a completely new experience that allows us to expand our view of the world and helps us to 'extrapolate' beyond our current experience.

As you have been reading this book, there may have been some explanations that you were already familiar with and you scanned through those sections quite quickly. In other parts you may have learnt some new information that filled a gap in your prior knowledge. It feels quite helpful and satisfying when that happens. You may also have come across ideas that you disagreed with, and these sections annoyed you. In at least a couple of places, I hope that you will have come across a concept that you have never seen before and found this very interesting and even exciting. Each reader experiences a different set of these eureka moments based on prior experiences and accumulated knowledge. It can't all be eureka moments, though – if it were, you may have struggled to reach this part of the book. Dealing with new information takes energy.

Even a single-cell microbe has a mechanistic way of assessing its surroundings, so this begs the question, does a cell have some basic level of consciousness? We may discover new methods that help us train intelligent machines so that they can learn from new information and even extrapolate more accurately, but a machine has no need for what humans would call consciousness, or emotions, and definitely no need for what we might call a 'soul'. And why doesn't it need a soul? Because an intelligent machine's directive, to gather information, originates from a method that we provide. Artificial intelligence is just a tool. It may be useful to create a method in our AI system that takes into account the rate of change of information, which could help the AI system learn from a new experience, but ultimately the goal of gathering and making use of information is set by humans and doesn't originate in a motive for survival.

As we have seen with the debate between neuroscientists, we don't yet know exactly how consciousness works in our own brain or whether it will ever be possible to break this down into a computational process. I am in the camp that says even if we did find a way to do this, it would just approximate consciousness and would not qualify in the way that we would describe consciousness for ourselves. Consciousness is a human – and perhaps a broader, biological – experience, and AI is delivered by a machine that we have trained. We get to decide what the machine will learn and how it learns, and therefore any method that we describe is by definition just a machine learning process (however complex and consciousness-like this method appears). It is certainly not conscious thought in the way we would describe this for ourselves; at best, it is a simulation of consciousness, analogous to the simulated black hole, which acts as expected without creating a 'real' gravitational field. The bigger question is one that relates to humans: are our conscious thoughts just a very complex biological machine process?

CONSCIOUS MACHINES

As we examine these ideas of consciousness, we can see that a computer should never be left to perform what we might consider in humans a conscious act. A malicious person or organization might try to train a robot how to kill. We already have so-called 'fire and forget' weapons systems that autonomously identify targets and are autonomously guided, but at the moment a human is still required to pull the trigger. I would strongly advocate, as most leading AI practitioners do, that the United Nations must outlaw the building of machines that can independently decide to take a person's life. The machine can provide information and knowledge that may help a human decide the correct action, but a human must make the final decision. They are ultimately responsible. However, even in the extreme case where a machine does perform a bad action, it is not the machine that is to blame, but the human who created it. An assault rifle can kill more accurately and efficiently than a pistol, but that does not remove the moral responsibility of the human using it – if anything, it makes human moral judgement more important by raising the stakes.

In his science fiction book *I, Robot*,[12] Isaac Asimov proposed three laws for robotics:

- First Law: A robot may not injure a human being or, through inaction, allow a human being to come to harm.

- Second Law: A robot must obey the orders given to it by human beings except where such orders would conflict with the First Law.

- Third Law: A robot must protect its own existence, but only if such protection does not conflict with the First or Second Law.

The problem here is that these propositions imply that the machine can reason and perform a humanistic conscious thought process that will allow it to make these difficult judgements. Unfortunately, this will *not* be possible. Any decision process will be the result of a machine learning method that has been described by a human. We can't offload responsibility to the machine; instead, the responsibility lies with the developer who created the method by which the machine learns from information, and so it is the developer of the machine learning method, or the organization or government that made the machine, who must be held to account. Our regulations and laws cannot be applied to the machine; they must instead apply to the developer and the entity that tries to use this weapon.

AI is a powerful tool that can become an incredible force for good, but we need to educate ourselves about its capabilities and ensure that we apply appropriate controls and regulations. We should worry less about AI trying to take control and running amok; instead it is the companies and humans who use AI, and the purposes to which it is applied, that we should understand and look to control. Appropriate controls around how companies and individuals develop and apply AI will be essential.

Educating ourselves about the underlying technology, its capabilities and its limitations is critical. Most issues that have occurred so far with AI have been driven by poor design and weak learning methods, or by the fact that most of today's AI systems focus on just a single goal or objective. In the popular documentary *The Social Dilemma*,[13] we see the way that social media platforms bring the full computing power of a large AI supercomputer to bear whenever you click on a social media link. The social media's AI system is focussed on getting you to click again. Rather than focussing on this single objective of getting you to spend more time on the platform, this documentary proposes that we perhaps need to train the machine to focus on a more complex, multidimensional objective such as 'time well spent by users'.

The key to building efficient AI systems will be to add more capability while at the same time ensuring that it is focussed on being a useful tool for humans. We need to increase the information capacity of AI systems so that they can become more accurate, and this means more compute and higher-performance memory. We also need to ensure that we do not introduce bias and that we control how the AI is used so that it doesn't manipulate the emotions of humans.

One potential trap that regulators can fall into is to make developers produce AI that is more understandable by making it follow some deductive process that tries to produce a definite answer. In some narrow applications this may be appropriate, but most applications will need a probabilistic answer and so we must be aware that the answer, although it may be accurate, will not always be 100 per cent correct. I also hear it suggested that we should 'adjust the algorithm', as though the system is somehow controlled step by step in a program. This won't work. Trying to make the AI work in some narrow way will make the result more brittle, less accurate, and prone to even more errors. Humans cannot fully explain how they make most decisions – the reality is we post-rationalize our actions. Most of our decision process is hidden in deep biological neural networks that make up our unconscious mind. Our human decisions are also heavily influenced by our emotions, by our values and by our experience. As roboticist Hans Peter Moravec wrote:

Encoded in the large, highly evolved sensory and motor portions of the human brain is a billion years of experience about the nature of the world and how to survive in it. The deliberate process we call reasoning is, I believe, the thinnest veneer of human thought, effective only because it is supported by this much older and much more powerful, though usually unconscious, sensorimotor knowledge. We are all prodigious Olympians in perceptual and motor areas, so good that we make the difficult look easy. Abstract thought, though, is a new trick, perhaps less than 100 thousand years old. We have not yet mastered it. It is not all that intrinsically difficult; it just seems so when we do it.[14]

Our human brain is incredible, and we are often completely unaware of how much is happening below the surface. Our conscious mind is still very much a mystery, and although we may eventually come up with machine learning methods that appear to mimic higher-level reasoning, we should never describe a machine as having conscious thought. More importantly, we must never believe that a machine, however powerful, has performed an independent conscious act. What we will have seen is the result of a machine learning method that has been described by a human. We must not fall into the trap of thinking that a machine has become sentient. It may mimic behaviours that appear to be very conscious, but it will just be following a machine learning method described by humans.

These are important points because we are on the cusp of building ultra-intelligent machines.

11

AN ULTRA-INTELLIGENT MACHINE

In 1968, when Stanley Kubrick was looking for a technical advisor to help him with his science fiction film *2001: A Space Odyssey*, he turned to a brilliant computer science professor called Jack Good. Good helped Kubrick create the profile for the film's infamous character the HAL 9000 computer. HAL is depicted as a sentient artificial general intelligence that controls all the systems of the spacecraft and is also meant to support the crew, but instead ends up taking charge. HAL is obviously science fiction, but it was rooted in Jack Good's belief that ultimately a machine could be constructed that would match or even exceed a human's capability.

Good believed that computers and their ultra-intelligent successors would deliver a benefit to humanity. In fact, the opening line of his seminal 1965 paper 'Speculations concerning the first ultra-intelligent machine' reads: 'The survival of man depends on the early construction of an ultra-intelligent machine.'[1]

It was in this paper that he also originated the idea of an 'intelligence explosion', something that has subsequently been called 'the singularity':

Let an ultra-intelligent machine be defined as a machine that can far surpass all the intellectual activities of any man however clever. Since the design of machines is one of these intellectual activities, an ultra-intelligent machine could design even better machines; there would then unquestionably be an 'intelligence explosion,' and the intelligence of man would be left far behind. Thus, the first ultra-intelligent machine is the last invention that man need ever make, provided that the machine is docile enough to tell us how to keep it under control.

American futurist Ray Kurzweil famously claimed in his 2005 book *The Singularity Is Near* that a singularity would occur by 2045.[2] I find this a slightly idealistic vision. As we will see, building a machine with an intelligence capacity that exceeds that of a human is now within reach. But I will also show that this machine could not, on its own, design and build even better machines. An ultra-intelligent machine may help humans design more powerful machines, but the machine will not achieve this on its own without human direction or human control. It will need help from humans – humans and machines would need to work together. This point is critical and is often missed. This is why we must understand how AI systems work, so that we can see what is possible but also recognize the limitations.

Good, a renowned British mathematician named Isadore Jacob Gudak by his Polish parents, had worked alongside Alan Turing at Bletchley Park in the UK during the Second World War. He came up with key breakthroughs that allowed the Enigma machine cyphers to be broken. Good subsequently worked with Max Newman and Tommy Flowers (who we met in Chapter 4) on the world's very first electronic computer and developed the algorithms that ran on Colossus, which unpicked the much more complex Lorenz codes that were used to encode messages between the Nazi High Command. After the war he joined Max Newman at Manchester University and together they worked to develop the 'Manchester Baby', the world's first stored-program computer.

As Jack Good identified, humans themselves have limitations. A typical human brain weighs around 1.4kg and consumes about

25 watts of power. Although very powerful, our brain still needs to be small enough to carry around. We need it to power our advanced vision system so that we can hunt more efficiently and to support our hearing and speech so that we can communicate effectively with others. Due to these requirements, our brain must fit inside our skull, and so it will be constrained in how large it can realistically grow.

Compared with an electronic computer, our brain uses slow components, with most of the neural-spiking pathways (which provide the communication in our brain) operating at just a few hundred hertz compared with the gigahertz in your laptop or smartphone processor. This makes our brain operations around 10 million times slower than a modern computer, and this is why computers can perform arithmetic much faster than you. However, your brain has many more parallel computing paths, and with roughly 86 billion neurons and over 100 trillion parameters[3] that store information, it is able to see patterns that computers find hard to recognize. Your brain is also far more power-efficient than a computer built from semiconductor integrated circuits. So, with all these differences, could we build an 'ultra-intelligent machine' similar to the machine that Jack Good first predicted?

BUILDING AN ULTRA-INTELLIGENT MACHINE

It has taken over 3,000 million years for humans to evolve from a single-cell organism to their current form. This evolutionary process, driven by a very gradual generation-to-generation learning process, has now resulted in a highly evolved and efficient intelligence system. When a human is born, they already have an optimized hierarchy of biological neural networks. Many of the structures are pre-trained, and humans can gather information that will allow them to quickly adapt to their environment.

As children grow, they are exposed to thousands of different objects and make hundreds of views of them on each encounter. Similarly, they hear several million words per year[4] and quickly build an understanding of language. The combination of our billions of neurons and the axons that connect them to create our trillions of synaptic connections allows us to capture knowledge. Researchers have not yet been able to build AI models of this scale.

We humans also keep all our information inside our processor: the brain. The information and memories don't need to be stored away in an external memory system. Information is available as new thoughts occur. Humans and animals just need to find the right connection (or association) in the maze of neural pathways. By contrast, an electronic computer needs to send information off to a memory chip, and as the problems become more complex, the access speed to this memory often creates a bottleneck. The very largest supercomputers have thousands of processor cores that operate at high speed, shuffling data backwards and forwards to external memory, whereas your brain has billions of incredibly small processing elements (your neurons) that all work in parallel at much slower speeds, but with your memories all held inside the brain.

No one is quite sure of the equivalent computing performance of our brain, but researchers have suggested that it is perhaps capable of 1 billion billion calculations per second – equal to a supercomputer with the performance of one Exaflop.[5] Whether that is a relevant measure is unclear because the amount of compute performance that an equivalent processor might need will be dramatically affected by how efficient the machine learning method is and how much information capacity the machine can efficiently access. Like the fuel consumption of a car, the amount of actual work that a computer can deliver will vary depending upon how efficient the processor and memory system are, and how well tuned they are to the AI application.

As a starting point, however, we can easily recognize that if a computer has a lower compute performance than our brain, or if the machine learning method is inefficient, the computer might still be able to do the job but it would just take longer than our brain. However, if the machine learning model is not able to deliver the same information capacity as a brain, or the computer running this massive model is unable to access the parameters fast enough to be useful, then it will not be able to match the capability of a human brain. The knowledge capacity of the machine and the efficiency of accessing this knowledge will be critical.

As we have also seen, a human brain comes pre-loaded with neural pathways and knowledge models that have been passed down through evolution. Humans have incredibly efficient learning methods that have evolved over hundreds of thousands of years. In contrast, each new artificial intelligence system needs to be trained, and this takes an enormous amount of compute and a huge amount of information. Typically, 100 thousand times or even 1 million times as much compute is required for the back-and-forth machine learning training process, compared with the computing performance needed for a single inference operation that provides an intelligent answer from a pre-trained model. Even a very powerful AI computer will be much slower at learning than a human.

An ultra-intelligent computer will also need much more computing performance and much more memory than will ever fit on a single semiconductor integrated circuit. To build an ultra-intelligent machine we will need lots of processor chips that are all connected together, and these connected processors will all need to work in concert.

In spite of all these challenges, we are approaching a time when it will be possible to build very large machines that connect thousands of specialist AI processors to deliver Exaflops of compute that can handle hundreds of trillions of parameters.[6] Such a machine might need tens of megawatts of power and would fill a warehouse-sized computing-data centre, but on paper could match or even exceed the knowledge capacity of a human brain. In comparison with a human brain, this machine would need roughly 1 million times more power and could only realistically be accessed via 'the cloud'. But such a machine would provide a platform for developing the next break-throughs in artificial intelligence and in a host of other fields.

The whole canon of medical journals could be read and under-stood by this machine to provide systems for doctors that would help them during diagnosis. The same would be true for our legal sys-tems. New metal alloys and composite materials could be investigated to build stronger and lighter vehicles. We might learn how to build more efficient machines for capturing renewable energy or use such a system to reach the holy grail of commercially viable nuclear fusion. Weather forecasts will become much more accurate. Also, safe, prac-tical, fully autonomous Level 5 vehicles might finally become possible at scale. These ultra-intelligent computer systems, used well, could also make the internet safer by helping to filter extreme material, as well as dealing with online bias.

Such a machine might also be used by us to explore new approaches to computing. It may help us build computers in new ways and with new materials that go far beyond the capability of silicon. This would open up the prospect for an intelligence explosion – not one driven independently by the machine itself, as many science fiction novels predict, but one driven by human curiosity and innovation. However, these next breakthroughs will not simply happen – they will take time and effort.

12

IS A SINGULARITY
EVENT POSSIBLE?

So, the question remains, could our ultra-intelligent machine become so powerful that it will start to reprogram itself and progressively grow more and more intelligent, leading to a so-called technological singularity? Is it even physically possible?

SILICON SCALING

Semiconductor integrated circuits, which we discussed in Chapter 5, have one fundamental disadvantage when compared with the biological systems that are used in the brain. Although semiconductor memories and processors have become enormously powerful, they are ultimately not the most efficient way to build a computer. Evolution has built far more efficient processing engines, and research has shown that modern computers are about 1 million times less efficient than biological systems.[1]

As we have already seen, in Chapter 11, a computer built from semiconductor integrated circuits that can match the knowledge capacity of a human brain might need tens of millions of watts compared with just 25 watts for our human brain. But even our highly evolved mammalian cortex is still roughly 100 times less power-efficient than a theoretically perfect biological machine. This computing power limit was described by an IBM Research scientist called Rolf Landauer in 1961. He believed that information is defined by physical systems such as strands of DNA, neurons and transistors. As a result, the limitations in the way cells, brains and semiconductors process information will be set by the laws of physics. He set out to measure these different systems and calculate whether there was a theoretical limit. What we now call Landauer's principle[2] transformed our understanding by highlighting the fundamental trade-off between information and physical systems. Landauer's principle shows that, for semiconductors, the minimum power that is required for a transistor to switch from a one to a zero will eventually hold back semiconductor-based computers. Ultimately, too much energy will be needed, and computers will not be able to keep shrinking in size while still getting better. Unfortunately, we are already approaching this limit.

We are close to building ultra-intelligent computers using advanced silicon processors that have been specifically designed and optimized for machine learning applications. These systems will continue to improve, but progress will be much slower than we have seen in the past. The revised prediction that Gordon Moore made back in 1975, that we would be able to double the number of transistors on a silicon chip every two years, is no longer true. It has taken nearly five years to achieve the most recent doubling in transistors and may take even longer for the next doubling. We need to adapt to this new reality. What we call Moore's law today was never the observation of a law of physics, it was always just a prediction of what could be possible if we worked hard and is a testament to the value of setting an ambitious goal.

Following Moore's original article, another research fellow working at the IBM Research labs, called Robert Dennard, did a more detailed study of what's known as the 'scaling' properties of transistors. The scaling theory of transistors that Dennard and his colleagues developed in the mid 1970s has underpinned progress in all semiconductor products.[3] Transistors in integrated circuits are built using a photographic-style photolithography process. An optical 'mask' (similar to an old-style photographic negative) is used to etch both the transistors and the metal connections into the surface of a silicon wafer (the base silicon platform on which semiconductor integrated circuits are built) through a complex set of chemical process steps called etching, deposition and implantation. Building a modern semiconductor IC chip requires approximately eight hundred of these separate process steps, with each needing to be incredibly precise. By gradually improving the manufacturing technology, from one generation to the next, it has been possible to shrink the size of connections drawn on the silicon chip, and as a result we have been able to make the transistors much smaller each time. As a comparison: a human hair is about 70 microns wide (a micron is 1 millionth of a metre); your eyes can see objects as small as about 40 microns; a red blood cell seen under a microscope is about 8 microns; the smallest human cells are about 4 microns. A modern semiconductor can have lines that are over a hundred times smaller than human cells.

What Dennard and the team observed was that if you can halve the size of the transistor (halving the area that it takes up on the silicon), then you can obviously fit two transistors where you could previously fit only one. Less obvious is the fact that because the transistor is smaller, its speed will also increase, because the electricity is travelling a shorter distance through the device. In fact, the speed increase is directly proportional to the decrease in the length of the transistor, and this works out to around a 40 per cent speed increase for a halving in area. However, the key discovery in Dennard's work was that the power that each of these smaller transistors consumes reduces in direct proportion to the reduction in area. When the transistor is halved in size, the power also halves. This means we get two transistors where we previously had one, where both are now up to 40 per cent faster and the power consumed by each is half of that of the previous generation. With each process-step improvement, you get two transistors for the price of one. That's why your mobile phone has kept on getting better while your battery still lasts just as long. The semiconductor industry has managed to continue driving this magical scaling since the 1960s and is why we have been able to build electronic products that are now 25 billion times more capable in just one lifetime.

With the transistors halving in size every two years, after twenty years you have reduced the size by 1,024 and after forty years by more than 1 million. The microprocessors we have today are about 1 million times more powerful than the early 6502 microprocessor that I used to build my first computer as a teenager. The early chips that I first used were in turn already 20,000 times more powerful than the first semiconductor integrated circuits built back in 1960.

Unfortunately, in 2007 we started to hit a hard physical limit. The voltage used for switching the transistor had been reduced so much that it could no longer go any lower. As a result, this power-scaling effect (known as Dennard scaling) slowed, and then stopped. The semiconductor industry has still been able to increase the number of transistors, but speeds have stopped increasing. If you were to keep on pushing up the speed of the semiconductor device, it would burn too much power. You will perhaps have noticed the impact of this with your own computer. The laptop you owned in the late 2000s had just one processor core and would operate at around 3 gigahertz. Today your laptop still clocks at around 3 gigahertz, but it will have four, eight or even sixteen processor cores, and in addition to this multicore central processing unit (CPU), your device will also include a specialized graphics processing unit (GPU) dedicated to graphics, high-performance networking and Wi-Fi.

A second challenge is now also emerging. The industry is starting to find it harder to keep reducing the size of the transistors. In the late 2000s the industry was able to 'draw' nearly 3 million transistors in 1 square millimetre of silicon. By 2021, the latest silicon process could fit around 135 million transistors in the same area of silicon, but it is taking longer, and costing more, to develop each new generation of semiconductor technology.

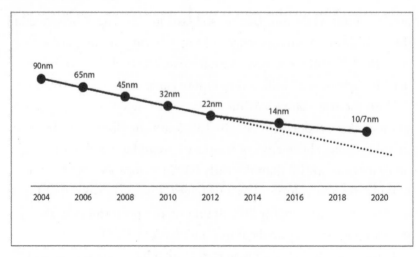

The dotted line shows Moore's law; the hard line shows actual scaling achieved

However, there are some new innovations coming that might help to mitigate this problem a little.

Over the last seventy years the focus in computers and microprocessors has been on increasing the rate at which we can execute instructions and on adding much larger amounts of external memory. Networks that connect computers together have also become important, especially for connecting to the internet and for connecting lots of computers together in a data centre. Clusters of computers inside data centres now work together on a single complex task, such as searching the Web or powering a large social media platform. Connecting processor cores inside a microprocessor chip, and connecting these processor chips together inside data centres, is going to become even more important in the future.

Nature gives us a clue as to what might work to connect all these separate processor cores together. Magnetic resonance imaging (MRI) information from human brains has shown us the highly complex three-dimensional network of connections that exists between neurons in the brain. Mapping this 'connectome' of the human brain is still a work in progress but we already understand some of the fundamental properties.

In studying the architecture of the human brain, neuroscientist Professor Edward Bullmore and his team at Cambridge University have discovered similar patterns repeating over and over again at different scales.[4] What they noted was a striking similarity to the much simpler structures found in the nematode worm. The nematode worm doesn't sound a very promising creature for research work on intelligence, but this extremely small and simple multicellular organism has allowed us to build a much richer understanding of how neurons and synapses work. Due to its simple structure, researchers have been able to fully map its neural connections. Its nervous system has been closely studied and this simple creature is one of the best understood of all life forms. What is striking is that the ratio of neurons to neural connections in the nematode worm appears to follow the same relationship as those we find in all brains, whether they are from humans, from animals, or even from insects.

In human brains, it is estimated that approximately 77 per cent of the volume is taken up with communication. As mammalian brain size increases, the volume devoted to communication increases faster than the volume devoted to functional neurons, and the ratio between the two approximates to something called a 'power law'.* The impact of exponential scaling is sometimes hard for us to visualize but it can be illustrated with a popular brain teaser:

In a lake, there is a single patch of lily pads. Every day, the patch doubles in size. If it takes 48 days for the lily pads to cover the entire lake, how long would it take for the lily pads to cover only half the lake?

---

* A power law is a mathematical relationship where an output varies in proportion to an input in such a way that the change in output will scale at an exponentially faster rate than the rate of change on the input. For example, the area of a square box will increase as a 'square law' (which is a power law of 2) when you vary the length of the sides of the box.

If you just quickly scan through this question, you might reply that the answer is 24 days, half the time it takes to cover the whole lake. But the correct answer is in fact 47 days. Each day the area covered by lily pads doubles in size, so from the 47th to the 48th day the lily pads go from covering half of the lake to covering all of it. On the 24th day the lily pads would still be almost invisible, taking up just 1/16,000,000 of the lake's surface. This logic leads us to a much-repeated quote about how technology develops, first stated in 1977 by the American scientist and president of the Institute for the Future, Roy Amara, and now known as Amara's law: 'We tend to overestimate the effect of a technology in the short run and underestimate the effect in the long run.'[5]

This power-law relationship in communication structures was first identified by E. F. Rent, another researcher at the IBM Research labs back in the 1960s, who first discovered what's now called Rentian scaling in electronic circuits.[6] As we look to build new computers in the future, we will need to take the effects of Rentian scaling into account. More and more area in our semiconductor ICs will need to be devoted to special forms of communication. The way processors talk to memory will also need to change, and the way memory is used to share information between processors will become very important. We can no longer rely on our processors just getting faster. Instead, it is extremely important that we develop new computer architectures that will allow us to continue increasing our capabilities.[7]

Software also needs to become much more complex. With new computers that have a very large number of parallel processing cores, a human can no longer directly write software for each core – this would just be too complicated. Instead, humans will need to describe a higher-level method for how these parallel systems should learn from information. This is how artificial intelligence works and so, by coincidence, silicon scaling has started to slow just at the point that artificial intelligence is starting to mature and needs new processor architectures. AI approaches are a practical way of developing the software that will be needed to run on these massively parallel computers.

We will still be able to build the first ultra-intelligent machines in the next few years by using these new techniques, but don't expect humanoid robots any time soon that have a low-power, ultra-intelligent silicon chip hidden inside. To drive progress towards even more intelligent machines, we will need to explore different materials. Silicon technology has taken sixty years to reach its current state of advanced development and semiconductor integrated circuits are the most complex product that humans build, but we are finally starting to hit the physical limits of these materials. A replacement is not going to appear overnight.

QUANTUM COMPUTERS

One candidate for more powerful computers comes from quantum technology. Quantum machines are extremely complex and hard to explain. They rely on quantum effects that happen at the subatomic level. Instead of using a switch to represent either a one or a zero, a quantum computer uses a quantum bit or qubit (pronounced 'cue-bit'). A qubit can exist as a one or a zero, but it can also exist in any state in between – and all at the same time. Think of it like a constantly spinning coin that could at any instant be either heads or tails – and not only either/or, but both at once. This quantum spin effect is exploited to perform computing operations. By firing a photon at a qubit we can control its behaviour by making it spin in a certain way so that it either holds information, changes the information, or reads the information out.

In a quantum machine you have groups of qubits all working together to perform calculations. By connecting lots of qubits together, complex high-dimensional spaces can be represented and explored. As an example, a large enough quantum machine could consider all possible chess moves – all 10^{43} of them – all at the same time. It is as if, rather than exploring a maze by turning left or right at each cross-roads, a quantum computer can turn 'left *and* right'. Programmed in the correct way, this quantum machine could suggest almost instantly what the correct move should be. It won't be much fun playing chess against a quantum machine – you wouldn't stand a chance.

Quantum computers aim to make use of 'entanglement', which causes qubits that otherwise appear to be behaving completely randomly to become correlated. This means that qubits can end up working (or spinning) together, and by exploiting this entanglement, computations can be performed.[8] Each qubit added to the system doubles the capability of the machine, and so if you can get large numbers of qubits working together then you could build a very powerful machine. Today, machines with a few hundred qubits have been built but experts in the industry say that we might need quantum computers with 1 million qubits or more before they will be able to do really useful work.[9] This is because there are a couple of snags that still need to be overcome, the biggest of which is noise. Quantum computers are incredibly fragile, and the slightest interaction with their surroundings will distort the results. Quantum computers very quickly become random noise generators, which means they give nonsense answers. Solving this noise problem and scaling to larger numbers of qubits is currently a challenge that many companies and researchers are working on.

Quantum machines are also not complete computers, in that they can only perform very specific types of work. To make them useful, they will need to work alongside silicon computers. The quantum machine may be able to propose the answer to a very complex question very quickly, but we won't know if this answer is correct (because of the possible effects of noise in the system). However, it is much easier to check an answer than to calculate one from scratch. So, in some important cases, if we can get quantum machines to work at scale, and we can get them to work alongside silicon computers as part of a total system, then they may help us to solve problems that are beyond the scope of today's computers. The challenge is that developing this technology to a point where it can become useful will still take some years.

For building AI systems, quantum machines may help in some respects. They can find patterns and associations in information much more quickly to help extract meaning, but they will not provide a wholesale replacement for our silicon machines. Classical and quantum machines will work together, leveraging the scalability of silicon computers and the specific problem-solving capabilities of quantum. Both have strengths and quantum computing is going to be one way for us to make AI systems go much, much faster.

MOLECULAR COMPUTERS

As we have seen, silicon computers will not be able to keep scaling at the rate that we have seen over the last sixty years. However, we will soon have ultra-intelligent machines that will unlock amazing new breakthroughs in artificial intelligence, hopefully accelerated by quantum machines. Perhaps we could use these artificial intelligences to work out how to build a new type of computer? As Jack Good said, 'an ultra-intelligent machine could [allow us to] design even better machines', and one area where an ultra-intelligent machine may help us is in molecular computing.[10]

To build molecular computers we need to learn much more about how biological systems work, and in particular how the molecules made up of a long chain of amino acids, called proteins, operate, and how these interact with DNA. It may become possible to find ways of building a computer using chemically controlled proteins that interact with artificial DNA structures to deliver logical switching circuits and computing engines. These molecular switches could potentially build much more efficient molecular computers.

This may sound speculative, but a number of researchers are already working on trying to apply the concepts learnt from DNA to build much denser forms of computer memory.[11] These DNA memory systems could store a huge amount of information in a very small space – in theory, a memory card the size of a tiny seed pod could store around 500,000 million million bytes (or 500 petabytes) of information. Early examples of these DNA memory systems have already been demonstrated in the laboratory.

However, it is not just memory systems that might become possible. By using a manufactured protein and causing it to change its folded state through a chemically controlled process, we could make molecular switches that are incredibly small and very energy-efficient. These molecular switches could be used to build logic circuits and to interact with storage elements in the form of DNA structures. Building deep artificial neural network structures could be possible. It may also become possible to scale these systems to build much larger molecular computers. Switching protein molecules will be very slow compared with switching silicon transistors, but their size and energy efficiency mean scaling could compensate and might allow very powerful machines to be built. Perhaps it will eventually become possible to build molecular computers that surpass the most powerful semiconductor machines but that operate with between 10,000 and 1 million times less energy. When this might happen is hard to say. Very early research is already taking place, but even small systems are perhaps three decades away. Ultra-intelligent computers may help us to accelerate this timeline.

This is all tremendously exciting. What we do know, however, is that there is no rapidly approaching 'intelligence explosion', and the much-written-about singularity is just not technically possible today. It will very likely remain impossible. Even quantum-accelerated machines and molecular computers will be constructed as a human tool and would be under our control. We will have no magic way of making these machines 'come to life'. Neither will they exhibit some innate self-conscious behaviour. These new machines will still need to be trained with information, and their knowledge will be limited by the information capacity and associated processing that such a machine can perform. Sentient artificial intelligence machines will not try to replace us, as in so much science fiction. The reality is much more interesting than that.

We have seen how we have been able to build powerful AI systems, and how they will continue to improve in the coming decades. New materials may deliver even more powerful machines, and with them, the next fundamental breakthroughs, though this will take time.

Today's silicon technology will not allow an intelligence explosion but may help us develop the next generation of computers. These new AI machines will continue to be directed by humans and can be controlled. However, we need to ensure that we remain up to date and informed about how these systems work. We must not become complacent and should work hard to anticipate the issues that need to be addressed.

*

Hopefully, we are starting to see that AI can augment our own human intelligence and that it can become a very useful tool – perhaps the most useful tool that humans have ever created. An AI revolution awaits, but do we understand the social and economic impact that this new technology will bring? AI has the potential to help us solve some of the most complex problems that humanity faces, but it could also create new problems. Do we know how to keep these incredibly powerful new AI tools under our control?

Looking at the profound changes that AI will deliver and the challenges that it can help us address, and how we can control it, forms the next part of our journey.

DO WE UNDERSTAND AI'S INCREDIBLE POTENTIAL?

There is not the slightest reason to doubt that brains are anything other than machines with enormous numbers of parts that work in perfect accord with physical laws . . . The serious problems come from our having had so little experience with machines of such complexity that we are not yet prepared to think effectively about them.

COMPUTER SCIENTIST AND AI PIONEER
MARVIN MINSKY, 1986[1]

13

AI AND THE ENVIRONMENT

Step into the control room of any major energy company and you will be confronted by massive screens that show all the energy sources and the major consumption points. A team of staff will be calmly trying to balance the supply to match consumer demand. They are working to try to make sure that power keeps flowing to your home. If this control room happens to be in Germany or, better still, in Denmark, then on a good day you will see well over 50 or 60 per cent of the energy coming from renewable sources, such as wind or solar power. However, in other countries around the world, the amount of electricity generated from renewables is usually much less. China and the US each get around 10 per cent from renewable sources today, and this roughly matches the world average. Countries in Europe are typically doing better, while most others are doing worse.

We think of the early Industrial Revolution as being a period of smog and pollution, and in 1800 annual worldwide carbon dioxide emissions from fossil-fuel combustion and industrial processes already stood at around 30 million tons per year. By 1950 our much more industrialized world economy emitted around 6 billion tons. In 2022, our supposedly post-industrial society produced over 36 billion tons (a 1,200-fold increase since 1800).[1] The number continues to rise as more countries become 'developed' economies.

About half of all carbon dioxide emissions will be dissolved in the oceans over a period of a few decades (which in turn is turning our oceans acidic). A further 30 per cent will be absorbed by vegetation over the course of a few centuries. But the remaining 20 per cent stays hanging about in the atmosphere for thousands of years. This means that even if we could halve the rate at which we are currently emitting carbon, we would still be outputting far too much and the total amount building up in the atmosphere would continue to increase rapidly. So, even halving carbon emissions would still mean that global warming continues, creating a massive problem for humanity.

As Bill Gates says, 'Most people now accept that to avoid a climate disaster, the world has to get to net-zero carbon emissions by 2050.'[2] Producing enough low-cost, carbon-free energy is one of the defining technical challenges of our time and one where artificial intelligence may help to provide some answers.

As we have seen, the human body consumes only about 100 watts per hour in energy, which is delivered through the food that we eat. However, humans need much more energy than this to power our advanced societies. In the United Kingdom, the per-capita energy consumption is around 4,000 watts per hour (or 4kW/hr). In China, the per-capita energy consumed is around 3kW/hr but is rising fast, and in the US the figure is much higher, at 9kW/hr. The world average is currently running at a little over 2kW/hr per person, but as more and more countries 'modernize', the amount of energy that we will need is going to keep rising. The total worldwide energy consumption of 18 billion kW/hr is likely to more than double and could soon reach 40 billion. It is naive to think that we can reduce energy consumption and maintain or improve our current standards of living. To maintain our modern society and ensure that poverty becomes a thing of the past, we will need to increase global energy production, but we must find ways to deliver this increasing demand in a zero-carbon form.

We also need to realize that energy is not the only source of carbon emissions. Many of our industrial processes also produce, as a by-product, carbon dioxide or other greenhouse gases, such as methane, nitrous oxide and perfluorochemicals (or PFCs), which are also driving up global temperatures.

Electricity accounts for around 20 to 30 per cent of our energy consumption, varying from country to country. Even in countries such as Germany, where a large proportion of electricity comes from solar power, or in Denmark, which on certain days is able to deliver all its electricity from wind, the problem remains that these renewable energy sources are completely dependent on the weather. To avoid interruptions to the electricity supply, energy companies need access to the most accurate weather forecasts to work out when back-up sources may be needed. Making the timing of each switchover as precise as possible is essential if we want to minimize carbon emissions.

Forecasting the weather is a complex problem, but new techniques are starting to combine AI-driven weather prediction with learning from historical weather cycles, while adding in real-time measurements from local weather stations. This information is also being combined with measurements from sensors located directly at the solar cells or on wind turbines as well as with satellite imagery. This much richer set of real-time information has allowed artificial-intelligence-driven forecasting to drive a 30 per cent improvement in the accuracy of weather prediction. In turn, this allows much more precise predictions for the actual power-generation potential from renewables, based on how hard the wind will blow and how brightly the sun will shine. The timing for when the grid needs to fall back on other sources of energy can now be much more precise. AI-driven improvements have extended this critical time window for preparing the appropriate back-up sources from minutes to hours.

As more information becomes available from smart meters and from other sources showing how we consume electricity, AI is also starting to be used to accurately predict upcoming power usage. This includes short-term peaks, such as during major events like a televised football match in the UK, when viewers rush to make a cup of tea at half-time, but also longer-term trends that allow renewables usage to be even more accurately managed. AI is also enabling predictive maintenance through close monitoring of sensors that are built into the equipment and into the power-distribution systems. Pre-emptive maintenance can be planned to coincide with times when demand is lower. This avoids having equipment fail that might otherwise cause energy blackouts or might force the use of older equipment, which in turn produces more carbon emissions. Monitoring the energy supply and level of energy demand also links to energy storage: AI is helping to predict when excess renewable energy will become available for capture and when smaller peaks can be filled using this stored energy, reducing the need to turn on carbon-emitting equipment.

Batteries and other long-duration energy storage technologies are becoming critical as part of the new renewables-driven energy grid. Researchers at the US's Lawrence Berkeley National Laboratory and the University of California, Berkeley, are using AI to investigate potential new materials for use in next-generation batteries.[3] The batteries required for energy storage in our power grid will be large and must be able to retain charge for a long period of time. They might be quite different to the small, light and portable batteries used in electric cars. Other storage approaches, such as using excess energy to compress a gas and then later using this pressure to generate energy, also show promise. Chemical processes that use spare energy from renewables to make a new chemical that can later generate energy are also being explored. These will help support long-term season-to-season energy storage. Another example is conversion of water to hydrogen, which can then be used as an alternative zero-carbon fuel. AI is being used to further all these approaches.

Electricity generation is still dominated by burning carbon-based fuels, accounting for around 25 per cent of emissions today. Ironically, as passenger vehicles and public transport switch to electric power, and as we transition other large industrial energy users away from coal and gas to using electricity, the percentage of carbon emissions that comes from energy suppliers could increase significantly. To avoid this, we must switch electricity generation to zero-carbon sources. Cheap carbon-free electricity could also be used to heat and cool our homes, as well as for cooking. But even if we can convert most of industry, transport and domestic energy requirements to using electricity, and we can continue the amazing increase in the use of wind- and sun-powered renewable sources, there is still a massive gap to fill before we can switch off coal- and gas-powered generators and achieve zero-carbon emissions. The world needs a safe, clean and continuously available energy source. Our own sun perhaps shows us what is possible. The British astrophysicist Arthur Eddington was the first to suggest in his 1926 paper 'Internal constitution of the stars' that stars, such as our own sun, draw their energy from nuclear fusion.[4]

Physicist Ernest Rutherford, with his famous experiment of 1934, showed that fusion was possible and reported that 'an enormous effect was produced'.[5] But it was not until the 1950s that researchers really started looking at the possibilities of replicating the process of nuclear fusion for power generation. Unlike nuclear fission, which is used in today's nuclear power stations, nuclear fusion is extremely safe because it cannot continue without containment, and so there is no way for a fusion reaction to end in a meltdown of the kind seen at Chernobyl or Fukushima. The amount of material required is also extremely small, and the 'burnt' fuel waste in a fusion reactor is an inert gas that exhibits very low background radiation levels. It's the perfect way to generate free and limitless energy, if only we could master how to do this as a commercially viable process.

To achieve fusion, hydrogen atoms are smashed together at incredibly high temperatures, creating a whirling plasma* that is hotter than the surface of our sun. In a star, the massive gravitational forces alone are enough to overcome the opposing atomic charges

---

* Plasma is an ionized state of matter, similar to a gas.

and push hydrogen atoms together. However, in a fusion system on Earth, large amounts of energy are required to achieve this reaction. Given enough force, the electrons become separated from the nuclei at extreme temperatures to form the whirling plasma. One method to contain this plasma is to use a doughnut-shaped device called a 'tokamak', which has incredibly powerful electromagnets that push the hydrogen atoms together to create a sustained fusion reaction. To extract energy, the plasma needs to be controlled so that it can fully heat up to between 100 million and 200 million degrees Celsius. It's extremely important that the electromagnets stop this incredibly hot plasma from touching the walls of the tokamak vessel. Otherwise, this could damage the reactor, but more importantly it would slow down or even stop the fusion reaction.

Experiments run at JET (the Joint European Torus fusion research centre in the UK) over the last few decades suggest that the complex science behind nuclear fusion power generation can work. Now, with some solid engineering execution, fusion could become a viable, safe and zero-carbon source of energy, able to supply energy to our power grids. Systems could come online by 2040. Based on this exciting research work, a new, globally supported joint fusion test facility is being built in France called the International Thermonuclear Experimental Reactor (ITER), with completion of this facility expected in 2025. This machine is expected to deliver ten times more energy as an output than the energy needed to start and control the fusion reaction. An experiment in the USA using a laser-controlled fusion approach has already released more energy than was needed to power these enormous lasers, achieving what is known as 'ignition' or energy gain. These breakthroughs are providing the next step in the journey towards a commercial technology.

However, one major problem remains, and that is how to control the containment and the shape of the plasma, which is a very unpredictable material. This process is critical to making fusion a viable energy source. A recent paper published in the journal *Nature* shows how a deep reinforcement learning artificial-intelligence system can be used to control the plasma inside a tokamak vessel.[6] An AI system was able to learn, through a self-supervised AI reinforcement learning method, how to control the powerful electric currents in the electromagnets in just the right way to maintain and adjust the plasma's shape. This problem is especially hard because the control process needs to continuously adapt to a constantly changing plasma, and even harder because there is only a limited number of sensors and the state of the plasma itself cannot be continuously measured. Despite this, the system, developed by DeepMind, was able to control the plasma and form it into different shapes. The system was then run on the tokamak in the Swiss Plasma Center at École Polytechnique Fédérale de Lausanne, where these simulation results were replicated in the real world. Others, such as the JET fusion facility in the UK, are working on similar projects, with the aim that AI can help us unlock the potential of nuclear fusion to deliver safe, clean and continuous energy that will move us to a zero-carbon-emission society.

Outside of energy generation, AI is also being used to improve manufacturing processes that today generate lots of carbon. Industries such as cement, steel, chemicals and plastics not only use large amounts of energy, but also generate carbon dioxide and other greenhouse gases as a by-product of their production process. AI is being used to search for new compounds that can improve these processes, reducing waste but most importantly reducing carbon emissions. Cement production accounts for about 8 per cent of global carbon emissions, but in a project undertaken by the social media company Meta and researchers from the University of Illinois at Urbana-Champaign in the USA, a new AI model was developed to explore different cementitious materials that bind the concrete, and to optimize the concrete mixtures for sustainability as well as strength. They were able to reduce carbon emissions by 40 per cent while increasing the concrete's strength.[7]

Another area where artificial intelligence is helping to drive improvements is in recycling, with AI-powered imaging systems now being used to analyse recycled materials and sort them much more accurately. These new approaches should have a major impact on plastic pollution and the growing amount of electronic waste that results from us upgrading our electronic devices so frequently, both of which are issues that must be addressed if we are to preserve our environment.

Continuing to damage our environment doesn't show much intelligence. We need help from artificial intelligence to solve this technological challenge. We should not expect our industrialized society to slow or reduce its insatiable demand for energy. Energy is fundamental to our wealth and prosperity; however, we must now rapidly turn to net-zero-carbon energy sources and time is not on our side. As Professor Vaclav Smil, the well-known authority on energy, points out, we cannot rely on AI to deliver a silver-bullet solution: 'The reality is that any sufficiently effective steps will be decidedly non-magical, gradual and costly.'[8] But in this most important challenge, we need all the help we can get.

14

AI IN EDUCATION

The Scottish comedian Billy Connolly tells a joke about how he hates algebra. He doesn't understand it, doesn't want to understand it, has never met anyone that uses it and never wants to meet someone that does. He complains that when his teacher introduced the topic at school, he must have been off sick on the day that he was supposed to learn how to add and multiply letters. He doesn't ever need to learn this new language because he has absolutely no intention of ever going to visit this strange place called Algebra.

For many this may sound familiar. How many children are there that fall behind in maths at school and miss out on a critical part of the syllabus? If you fail to grasp basic algebra then calculus becomes impossible, partial differential equations remain a mystery, and linear algebra, which is crucial for understanding the maths behind artificial intelligence, is well out of reach. But it is not just complex subjects like maths where people struggle – even basic literacy remains a challenge.

Even though over 90 per cent of the world's population have access to electricity and are also in reach of a mobile internet signal, only about 50 per cent of all the world's young adults complete secondary-level education. Only 7 per cent go on to earn the equivalent of a Bachelor degree. Not only do these children miss out on algebra but, according to the United Nations, over 600 million children and adolescents worldwide (55 per cent of the total) are still not achieving minimum proficiency levels in reading, writing and mathematics.[1]

Despite years of steady growth in school enrolment, non-proficiency rates remain high.

The statistics for 1940s America were very similar, with only 40 per cent of people attending high school and only 6 per cent earning a Bachelor degree. Just four generations later, the situation looks very different: 88 per cent of young people are now completing high school and 60 per cent attain a higher education qualification from a college or university. However, more than half of Americans (54 per cent) between the ages of 16 and 74 still have reading skills that are below the equivalent of Sixth Grade level (equal to an average 11- or 12-year-old). Can AI help to solve these educational challenges?

The recent pandemic forced more than a billion students around the world to try to learn via a screen at home, with varying levels of success. However, a new approach called 'personalized learning' is emerging, which is set to change this. This approach combines AI-powered interactive learning on a computer together with support from teachers. The idea of AI tuition is that students can learn at their own pace, and so, rather than falling behind and feeling that they are no good at a particular subject, they can spend a bit more time to catch up, or they can go back and repeat something. Students who find the topic a bit easier can run ahead and avoid getting bored. The teachers can then focus on coaching and on providing one-to-one support. Personalized learning systems take advantage of AI to create a highly interactive experience for students, and they work alongside teachers to really improve education.

China is leading the way with these new solutions. Even First Graders are learning to write complex Chinese characters, with AI-powered computer vision and speech synthesis helping to deliver the lesson alongside teachers. AI image-recognition systems look at each Chinese character that the student writes and provide an immediate assessment. Speech synthesis and online videos give these six-year-olds real-time instructions and allow them to progress at their own pace. The young students can progress quickly and they don't get bored, but if they need more help the system will slow down and allow them to repeat exercises. It even tries different teaching styles and approaches until they can master the subject. Human teachers are able to focus on providing targeted support to the students that really need it, reducing the amount of time that they previously spent developing and marking tests.

The results are striking. With fun animated characters helping to keep young students' attention and human teachers stepping in as required, pupils make rapid progress. AI teaching systems can even monitor the students' facial expressions, and will learn when the students have understood or can be seen to be struggling. With some of these systems, correct-answer rates have increased from 50 per cent to 80 per cent. But the AI systems are also able to learn what works best for different types of students, with options for more visual, auditory and written lessons, or ways to let them try things out physically. With the software capturing information from how the students use the system, the AI can be continuously improved.

One start-up company that is participating in China's experiment in AI-powered learning is Squirrel AI. They are trying to address what they see as the 'lack of personalized attention in traditional classrooms' and an 'unequal distribution of educational opportunities'. Their personalized tutoring services are helping students across a range of age groups. In one example, a thirteen-year-old student who received tutoring support from this AI system was able to raise their maths test scores from 50 per cent to 62.5 per cent after just one semester, and achieved 85 per cent in their final middle school exam two years later. 'I used to think math was terrifying,' they said. 'But through tutoring, I realized it really isn't that hard. It helped me take the first step down a different path.'[2]

But education doesn't need to stop as you get older, as a ninety-two-year-old in the UK recently proved by sitting a secondary school maths exam. AI-driven teaching is not age specific and can be applied across the whole educational spectrum, from degree-level courses to workplace learning. Often, older people are not willing to risk being embarrassed in a schoolroom, so a personalized system provides a much safer place. Mature students who want to learn can take an AI-enabled course on their own with help from a virtual tutor and can learn new skills at any time. AI-powered approaches can be much more interactive than just watching a video – you can progress at your own pace and avoid the stigma that can be associated with an in-person adult literacy class.

AI also significantly reduces the cost of education and can make great learning platforms available to many more people. Accurate language translation is possible, which means that courses developed in Europe, the USA or in China could easily be translated and shared across the globe to improve education in even the most remote spots. As an example, one ed-tech start-up based in Nairobi, Kenya, is using mobile phones and artificial-intelligence systems to analyse the lessons of students, and then delivers new personalized content. The system also provides updates on the students' progress to the teacher, so that they can step in and help as required. The content and assessments are delivered through very basic mobile text messages, which is cheap and broadly available. This makes the whole system easily accessible to students across Africa, where over 80 per cent of students are currently failing to reach minimum proficiency levels (according to UN research).[3]

What is striking is how new, large language model-based systems and other knowledge-based tools could really help in the classroom. In one recent example, ChatGPT was tried at an intense software development training session held by the Alperovitch Institute, a group that focusses on cyber security and is based in Washington DC.[4] What they found was that having ChatGPT open and available allowed students to find answers to basic questions, getting them up to speed on issues that they were not familiar with. It saved the class from being disrupted by questions that only a few students needed the answer to. It also helped people to quickly find additional materials on the subjects that were being discussed by the tutor, deepening their knowledge. But they also found that ChatGPT was able to suggest answers, show useful coding techniques and compare students' answers with ChatGPT's answers. This simple experiment was able to turbo-charge the teaching environment and left everyone clear that AI will transform education.

AI-powered education systems will help teachers by covering many of the distracting routine tasks – lesson planning, creating assessments, and even the basic sharing of facts – that take teachers away from actual teaching. Instead, teachers will be able to focus on supporting and inspiring students to learn, which can then become a lifelong skill for the students. Students can use AI to build up their knowledge and it allows them to learn at their own pace or helps them explore related areas that broaden their knowledge. Too much of education today is focussed on just remembering facts that allow students to pass an exam. Often, students will quickly forget this information unless they can apply it in real-world situations.

Education is an area that the Bill & Melinda Gates Foundation focusses on, and the Microsoft founder Bill Gates has said that AI in education 'will know your interests and your learning style so it can tailor content that will keep you engaged. It will measure your understanding, notice when you're losing interest, and understand what kind of motivation you respond to. It will give immediate feedback.'[5]

My own view is that AI will help us transform the world's education systems by shifting the focus from the basic 3Rs (Reading, wRiting and aRithmetic) to instead developing the 3Cs: Curiosity, Critical thinking and Creativity. AI can help us all build lifelong learning skills and augment our human intelligence.

15

THE AI HEALTHCARE REVOLUTION

If you had been born in the year 1900, you might well have lived to just thirty-one years, which was the average at that time. Modern science and breakthroughs in healthcare have now extended this global average life expectancy to over seventy-three years, giving people more than four extra decades of useful life. But the recent Covid pandemic has also made us all realize just how fragile our overworked healthcare system has become. The good news is that a revolution in healthcare is coming. Indeed, our health is perhaps the area that will be most positively affected by new technologies.

AI can bring new breakthroughs in diagnostics, in drug discovery and in improvements to our personal health. But it also has the potential to significantly improve one other very important aspect of our healthcare systems that has been in steady decline over recent years. As Eric Topol says in his book *Deep Medicine*, 'The greatest opportunity offered by AI is not reducing errors or workloads, or even curing cancer: it is the opportunity to restore the precious and time-honoured connection and trust, the human touch, between patients and doctors.'[1] In most developed countries, our overworked healthcare services are struggling to cope. Many believe that AI is the only solution. It will allow us to move, at last, from a system that treats people who are sick to a health system that keeps us healthy.

DIAGNOSTICS

When we get sick, we put ourselves in the hands of our doctor, who acts as a detective, searching for clues that will let them uncover what it is that is ailing us. The doctor starts with only the very limited amount of information that we can share about what feels wrong – maybe some visual signs, plus perhaps relevant historical information from our medical records. From these very limited clues, our doctor will need to make decisions about what tests could help to provide more information. Artificial intelligence can help the doctor to dig deeper into the information and augment their enormous skill, speeding up the diagnosis.

Using information from blood tests and DNA swabs, AI can open up a whole new world of knowledge about what may be happening inside our bodies. Traditional tests are powerful, but AI is helping to build a much richer picture of the cells in our bodies, and how they operate.

So much more information can now be uncovered. New sensors are being combined with AI systems that can search through the complex sequences of information in our blood, looking for patterns that might uncover an obscure health condition. This level of analysis, or the speed at which results can be provided, would just not have been possible using previous lab-testing approaches. This rich source of new information also builds over time, as more tests provide more information. New connections between complex pieces of information are being uncovered that can help identify new patterns and clues. The amount of information that can now be analysed, and the obscure connections that this will uncover, will transform disease diagnosis.

Molecular analysis company Oxford Nanopore, based in Oxford, UK, has developed a new type of sensor that they combine with a powerful AI system to examine infectious diseases, to help match organs to patients in major transplants and to provide information on reproductive health. The company are using this information to build a clearer picture of how our bodies work so that illnesses can be diagnosed earlier. Their technology is able to uncover connections that even the most experienced clinicians would struggle to find, and they are able to do this much faster.

But sometimes doctors also need physical information, and new approaches to three-dimensional body scans using computed tomography (CT) scans, which use X-rays, or magnetic resonance imaging (MRI) using radio waves, are building very detailed physical pictures of our bodies. Far from making clinical radiologists and oncologists redundant, AI is helping to improve workflows and uncover new insights. AI can optimize the radiology and oncology processes and is providing much more detailed clinical information. These techniques are already being used to detect possible cancers and to look deep inside our organs by combining different image segments of organ scans to build up a full 3D view. The time to 'read' and compare scans is being drastically reduced, and this is delivering earlier detection of disease, improving diagnostic accuracy and providing a more personalized result.

A key issue in CT scans is the overlaying and comparison of images that have been taken at different times. This process is important for tracking changes in a patient's condition, but often the images will be subtly different and can easily be affected by changes in body shape that are caused by breathing during the CT scanning process or by weight changes in the patient between scans. The AI systems are able to identify different elements in the image, including bone, organs and tissue, and can use these to align the images not just in two dimensions but in three. This then allows the growth or shrinking in size of anomalous tissue to be accurately measured from one scan to the next. The AI systems help radiologists to focus on what is really changing and can dramatically speed up the diagnostic process, while providing much more accurate results.

US hospitals alone admitted 33 million patients in 2020, and just the sheer logistics of handling this many patients, with all their unique ailments, is a particularly challenging diagnostics problem. In healthcare, time is money, and being able to treat patients efficiently and successfully not only leads to better outcomes, but also makes hospitals much more efficient and able to handle more patients. This new level of efficiency ensures that much more time is available for the most challenging cases, improving the level of care available for *real* emergencies. New innovations in AI healthcare solutions are already helping to streamline the patient experience. With the right software, hospital staff can start to process billions of pieces of information, allowing them to shift away from caring for the sick to helping to pro-actively prevent sickness. This in turn leads to better health and less health-related conditions. It is becoming possible to start rebuilding a more human-centric healthcare system.

In India this situation is starkly illustrated. There is a marked dif-ference in life expectancy for people living in urban centres, who have easy access to doctors, compared with those in rural areas. According to data from the Registrar General & Census Commissioner of India, life expectancy for women is around seventy-four years in urban areas but reduces to just sixty-nine in rural communities.[2] The same figures for men are seventy-one years, reducing to just sixty-six. India has a huge population, and proportionally fewer qualified medical profes-sionals. Adding to this challenge is that India has twenty-two different official languages, and so providing even the most basic healthcare hotline is challenging. Matching a person from a rural area with enough medical professionals that have the correct language skills is a challenge that the Indian health service is struggling to address, and they see AI as offering a solution. Not only could it translate the full range of regional languages, it could also provide a voice-activated chatbot that can be used to gather information from rural patients via a simple mobile phone conversation. An AI system could then provide simple medical advice for very basic cases and perform a first level of triage, steering patients towards an appropriate clinic or hospital for a much more in-depth consultation.

Healthcare already generates huge amounts of highly relevant and sensitive information. Connecting all this incredibly important information can really drive healthcare knowledge and speed up diagnosis. If this critical information is used effectively to train new AI systems, then outcomes can improve. However, this highly sensitive information must also be protected and used appropriately, with strong regulation needed to safeguard people's private data (a subject we'll explore in more depth later on). But to highlight the point, in 2017 London's Royal Free hospital was found by the Information Commissioner's Office to have mishandled patient records when it handed over personal data of 1.6 million patients to DeepMind, a Google subsidiary. Google used the data to test an app that was trying to identify acute kidney injuries, which, if identified early, could address the 25 per cent of preventable deaths caused by kidney issues. To highlight how this case shines a spotlight on how patient information must be used correctly, the lawyer leading the legal claim stated: 'It should provide some much-needed clarity as to the proper parameters in which technology companies can be allowed to access and make use of private health information.'[3]

We must keep patient information confidential, but we should also recognize the potential for AI to allow doctors to uncover new patterns that will help them make a suitable diagnosis even from the smallest clues.

DRUG DISCOVERY

Putting a single new drug through a full clinical trial costs on average $1.3 billion (approximately £1 billion) and these costs have been rising, driven by the increasing complexity of laboratory research and approvals testing. Today only 10 per cent of the new drugs in development successfully make it through this complex process and become available on the market.[4] These skyrocketing costs are holding back the whole pharmaceutical industry, and AI perhaps holds the key to addressing this fundamental healthcare problem.

We still understand so little about our cells and the ways in which they interact. We are starting to learn more and more about the importance of proteins and the interactions of bacteria and viruses. There are exciting developments going on in a number of related 'omics': the study of genes (genomics), RNAs (transcriptomics), which translates genetic information into activities in the cell, proteins (proteomics) and metabolites (metabolomics), a substance in cells that produces energy as the end product of metabolism. AI is now helping us build a much deeper understanding of this whole complex area and could transform the way in which drugs are developed and tested.

When we are infected by a virus such as the common cold, the virus uses our cells to reproduce and start living off our bodies. Our immune system releases antibodies to try to kill the virus, but viruses have been honed by billions of years of evolution and they will sometimes find ways to use our own antibodies against us by triggering an antibody response that lasts well beyond the initial infection. Understanding this microscopic warfare and working on drugs that can operate at this cellular level is a new area of research that has started to be unlocked by advanced computer simulations. AI systems are helping us to decode the complexity of the intelligent cellular systems that make up our bodies. They are able to predict the three-dimensional structure of proteins and this is an area where the most amazing progress has been made – as we have already seen with our description of the AlphaFold system, in Chapter 5.

During the drug discovery process, a greater understanding of proteins helps researchers in lots of ways. This includes analysing novel drug targets, understanding the mechanism of diseases, optimizing the chemical structure of a drug based on the structure of the protein target, and helping to identify leading chemical compounds that show promise in the treatment of a disease. This new area of AI-driven drug discovery is still in the very early stages but could end up having the most profound effects. Think of it like a computer game, where the gaming environment is a simulation of our cells that the AI system is trying to understand. The AI system can learn how to find exactly the right chemical molecule that could attach to a protein in the affected cells and deliver a drug that will kill a virus, stop cancer from growing, or block bacteria that do us harm.

The UK AI drug discovery company Exscientia was the first company to have an AI-designed molecule enter clinical trials. They have shown how their approach was able to identify drug candidates from molecules based on their biological effects. This difficult job is normally reserved for scientific experts who have built up an enormous understanding of chemical structure and biology. But their AI system is able to explore many more permutations to test the suitability of molecules and identify whether any hold promise for a new drug. In tests comparing the results from this AI system with ten human experts, the AI was able to match the results of the very best expert, but they were achieved in milliseconds rather than taking several hours for the scientists to produce.[5]

One problem that Exscientia is focussed on is finding molecules that can hit two targets at once, with special molecules described as 'bispecific'. As their CEO Andrew Hopkins says, 'if you add the need to hit two targets, rather than just one, then the complexity goes up enormously. Finding a drug is like finding a needle in the haystack, but finding a bispecific small molecule is more like finding the needle on the farm.'[6]

This may sound like an impossible task, but it's already happening. Computational methods have actually been used for small-molecule drug design since the 1970s, with researchers focussing on understanding the relationship between the molecules' structural characteristics and their biological activity. The fact that digital information has been built up over time for these small molecules makes this an active area for applying new AI methods. The formation of small molecules is closely linked to their chemical structure, making it possible to build models that can explore these chemical associations too. Artificial intelligence can be used to predict the properties of different small-molecule candidates, dramatically increasing the certainty of success in clinical trials.

In 2019 researchers from Insilico Medicine, a Hong Kong start-up, published an article in the magazine *Nature* showing that they had succeeded in using AI to design a new molecule that could potentially be used as an inhibitor in fibrosis (the thickening or scarring of tissue) and other diseases.[7] They achieved this incredible result in just twenty-one days and were then able to validate this in only twenty-four more days.

AI can now be used to search across a vast quantity of existing small-molecule information to learn patterns and associations that are far too subtle or complex for a human to find. This is happening at a speed that was not previously possible. This new AI understanding of how these small-molecule chemicals interact with protein targets lets scientists steer their research. Precious lab-testing time and clinical trials can now be much more focussed on investigating only the most promising candidates, accelerating new discoveries and new treatments. The potential is enormous, not just for humans but even for understanding health conditions in other species. Your cat could soon be made healthier, too.

Understanding more about DNA is another area of active research. DNA is an encoding of a sequence of information, just like language is. Since AI is now capable of translating from one language (one sequence) to another, variations of these sequence-to-sequence translation methods are now being applied to understand DNA. AI opens up the prospect of finding identifiers in our DNA that could predict genetic conditions and provide early warning signs for cancers or other complex issues. A simple blood sample or DNA swab could allow an AI system to discover much more information about us and our future health, allowing us to take early precautionary steps – another way of keeping healthy rather than waiting to get sick.

Rare diseases are also, at long last, starting to get some attention. The cost of drug discovery is so high that, up to now, pharmaceutical companies could not afford to focus on rarer conditions. However,

Healx, a start-up company based in the active technology hub of Cambridge, UK, is using AI to tackle this important problem. They have developed an AI-powered biomedical knowledge graph that captures rare-disease information. Knowledge graphs take data from many sources and find context to turn this data into associated sets of information, from which knowledge can be gathered. They are using this approach to find associations between rare diseases and existing drugs that are already approved for use in curing other medical conditions. Approved drugs have already been through an enormously rigorous testing process and so we know they are safe for use on humans. Healx have built their AI system to help identify novel connections between these known chemical entities and rare diseases that cause severe issues for the unfortunate people affected. These connections can translate into new treatments that are then able to move rapidly to trial as therapies. Because they are using existing approved drugs, the trials are happening much faster, and if the drug is found to be effective, it can be scaled rapidly into production for this new application. It's a nice idea, but the most amazing part is that it works: Healx have already received a US Food and Drug Administration (FDA) approval for a Phase 2 clinical study to treat Fragile X syndrome, the leading genetic cause of learning difficulties in young people; the condition causes developmental delays and learning difficulties, plus social and behavioural problems. Fragile X is easily diagnosed in young children by performing a DNA test on a blood sample. Unfortunately, there have not been any effective drugs – until perhaps now. Many more advances are in the pipeline.

Reducing the time and cost of drug discovery has the potential to greatly expand access to treatments. In the past, rare conditions were often overlooked because the cost of drug development would completely outweigh the commercial return. In the cases where specialist drugs were developed, healthcare providers would sometimes limit distribution to keep prices high, helping to cover the expensive cost of development.

Since 2018, when Martin Shkreli, the former CEO of Turing Pharmaceuticals, was sentenced to seven years in prison for fraud, the whole debate around the opaque pricing model used by the drug industry has come under the spotlight. His firm had acquired the rights to a sixty-year-old drug called Daraprim and then raised the price from an already expensive $13.50 per pill to $750 each. Daraprim is an antiparasitic that is used to treat a parasite infection called toxoplasmosis that affects the body, brain or eye, and the drug is also used for people with HIV.

As companies like Healx and Exscientia are showing, there is an alternative route. The improvements that AI is starting to deliver, in all aspects of drug discovery, could have a profound effect. The way in which AI will transform drug discovery and make treatments accessible to all, everywhere on the planet, is only just starting to become apparent. We should expect massive progress over the next few years.

PERSONAL HEALTH

Our ageing populations are pushing healthcare provision around the world to the brink of systemic failure. These critical services are struggling to cope, with growing waiting lists, increasing incidents of chronic conditions, and major challenges in hiring skilled staff. AI may be the only realistic solution.

People today are having fewer children, and as childbirth rates decrease and the elderly live longer, so the average age of our population is increasing while at the same time the total population level is starting to plateau and will then decline. The effect of this is already visible today in Japan, where in September 2021 it was reported that 29 per cent of the population is aged sixty-five or older, with a record 80,000 centenarians. The birth rate in Japan is just 1.36 children for every woman, compared to 1.76 in the USA, both far below the 2.1 childbirth level required to sustain a population. By 2036, people aged sixty-five and over will represent well over one third of the population in Japan, but Europe and China are not far behind, with 20 per cent and 12 per cent of people now aged over sixty-five; in the UK the population is already ageing rapidly, driven by low childbirth rates, and in the case of China, by a one-child policy that has only recently been revoked. Three hundred and thirty million Chinese are expected to be aged over sixty-five by 2050, and childbirth rates in China have declined from over six per woman in the 1960s to around 1.3 today. As we can see, the percentage of older people in developed nations is growing rapidly, and it is projected that the combined senior and geriatric population in the world overall will reach 2 billion by 2050.

Staying healthy for longer will be key if we are to avoid a health system that becomes overloaded by older people suffering from chronic conditions. Fitness tracking and personalized fitness programmes are being powered by AI systems that help us understand so much more about our own well-being. Smartwatches are providing constant information about our health by measuring our exercise and heart rates, and even include blood-flow monitoring. New technology is starting to appear that lets people with diabetes monitor their blood sugar level from their wrist, with the same technology able to provide feedback on the state of your metabolism. It is also able to check our sleep patterns, and you can now see whether that extra glass of wine helped you to fall asleep – or perhaps, as we all suspected, caused dehydration and disrupted your healthy sleep pattern.

Encouragement to engage in more fitness activities and to eat healthy foods will all help as we age. As people get older, they want to remain self-sufficient and lead independent, fulfilled lives. It's possible to make healthcare more personal and efficient for our ageing population, giving them a higher quality of life and the ability (in most cases) to avoid institutional care. Mobile devices can provide constant monitoring by keeping track of people and taking regular measurements. We could also add a few internet-connected sensors in the home that serve a dual purpose of providing temperature- and security-system control. These sensors could alert us when a medical emergency (such as a bad fall) occurs.

It's easy to collect this kind of health information; the challenge is what to do with it. One option is to set up a private website that displays this information in a dashboard, so that you can monitor this from time to time to check that all is well with your loved one. But while this might be useful, it is a little intrusive and may not be the best solution. Fortunately, there are other options. A few specialist AI companies have been developing systems that can learn and make predictions from smartphone and other sensors, and from these predictions can monitor people's condition. More importantly, the system can use sensor information to understand the context of the user – if they're in their car, on a bike, hurrying for a bus or at home asleep.

Being context-aware allows more accurate predictions to be made from 'regular' health information. For example, we may find that a person's temperature and heart rate are high – if we know that they went out for a bike ride, we could infer that everything is normal. We wouldn't be able to make this inference without context. It should be possible for the system to know when to get a care worker involved so that preventative steps can be taken. These types of AI-powered systems have huge potential to provide holistic healthcare that can really improve people's lives, while respecting privacy and supporting independence. The key challenge is to understand the context, and this includes monitoring how the information changes over time. This type of temporal AI model can uncover insights from changes that happen rapidly or can identify subtle effects that happen over a much longer time period.

To help its ageing population, the government in Japan has even been funding the development of elderly-care robots to help deal with the lack of care workers and to stem loneliness. Furry robot seals and robot dogs have been developed that engage in conversation, and humanoid robots assist with mobility exercises. This may sound scary, but according to the Japanese indigenous religion, Shinto, spirits (*kami*) must be attributed not just to people, but also to animals, mountains, kitchen knives and even pencils. As a result, the Japanese typically don't see a distinction between humans, animals, plants and objects – including robots. Most elderly Japanese see nothing strange in a robot that demonstrates human or animal-like behaviours – it is just *kami*.

We don't all simply want to live as long as possible; we want to lead healthy, independent lifestyles. Soon, 2 billion elderly people will be looking for solutions that will help them live fulfilling lives. But AI-powered systems will need to be extremely intelligent if they are to advise and support in a way that is thoughtful and non-intrusive, while respecting our privacy and only intervening when really necessary.

The elderly aren't the only people who could benefit from these AI-powered life-enhancing systems. Patients with ongoing chronic conditions can have closer monitoring and remote support, with changes in their conditions becoming much more visible to their carer. Healthcare workers will be able to use their time more efficiently, focussing on proactive and needs-based treatment. For health workers, it is much more rewarding to help people lead enjoyable, independent lives than having to deal with health problems that destroy a person's quality of life.

Companies such as Babylon Health, who describe themselves as a digital-first healthcare provider, are already using AI to provide a new approach. By offering a service through mobile devices, they deliver proactive healthcare, switching away from the current reactive system. Their AI service helps to provide health tips and feedback on symptoms, as well as delivering future health predictions. When necessary, it will connect patients directly to a human doctor through the patient's mobile device. This combined AI- and human-based approach is making healthcare much more convenient and available for all.

I recently attended a workshop where social enterprises were focussing on delivering AI solutions to the growing concern of mental health. Students at college seem particularly vulnerable but mental health has also become a more prevalent issue among those who work in stressful roles. Employers and colleges are searching for solutions that can help. AI systems can provide a safe environment for people to openly share information about how they feel. Clinically validated technology can then compare people's inputs against AI models built from hundreds of thousands of pieces of relevant information so that patients can be matched to the correct specialist for either an in-person or remote appointment. These types of personalized systems are already making a real difference in keeping people well.

Happily, health insurance companies, governments and taxpayers all stand to benefit. As the population ages and as people experience more stress, public health systems and institutions are having to bear the cost. But AI systems could improve and support our healthcare systems, helping to avoid the rising cost of institutional care and adding significant value to the economy. Healthcare is a public good, and artificial intelligence will make it possible to improve outcomes for all.

WORLD HEALTH

In some parts of the world, people's faces will light up when you ask them about their toilet. For people who have lived without running water or access to clean sanitation, the arrival of this most basic of facilities is completely life-changing. According to Unicef, 3.6 billion people around the world (nearly half) do not have access to safely managed sanitation facilities. Open latrines flood when it rains, and attract flies and insects that then spread waterborne diseases. Having a simple covered latrine, with a toilet bowl, installed over the hole to create a seal that keeps out flies, insects and odours can transform people's lives.

Meanwhile, nearly 1 billion people also rely on public water taps that are a focal point for many villages in Africa and Asia. Rather than having to walk two hours each day to get drinking water, this life-giving liquid can now be found just a few steps away. Unfortunately, these water and sanitation resources will often start to break after only a few years of service, and roughly 25 per cent of these essential services are not functioning at any given point in time, which has a massive impact on the communities that they serve. These services obviously rely on physical solutions, most of which are very low-tech. But AI has an important role to play in ensuring that the resulting services are maintained, managed efficiently and can, more rapidly, become available to all.

In one AI initiative, led by the Global Water Challenge,[8] over 500,000 different pieces of information were analysed to predict points of weakness across the water network in thirteen African countries. This analysis then provided predictions about where preventative maintenance was required – information that was passed to the local water network operators in these countries. This project has resulted in reduced costs and a much more reliable supply, and is also allowing the water networks across these regions to be expanded more quickly.

Keeping people fed is another problem that needs a lot of low-tech physical solutions. Improving our diet and monitoring our level of healthy exercise are all First World problems. But over 2 billion people (a quarter of the people on our planet) are either not getting enough food or are unable to eat a healthy balanced diet on a regular basis. Just providing the basics of food, water and sanitation for all is critical so that everyone can enjoy a healthy life.

A recent McKinsey management consultancy study[9] showed that over 60 per cent of people in sub-Saharan Africa are smallholder farmers, and this drives over 23 per cent of GDP for the region. But much more productivity is possible, and the same report highlighted that two or three times more cereals and grains could be produced, with similar gains possible in horticulture and livestock farming.

AI is already being used in various ways to improve farming

efficiency, including crop-yield improvements, targeted irrigation, soil content analysis, crop monitoring and better crop establishment. But 50 per cent of all droughts across the globe affect only Africa, and the frequency of these events is increasing because of climate change. AI is already being used to predict droughts and could help to alleviate their worst impacts by offering an early warning so that food and essential supplies can be provided before they run out. Another key application is in the highly targeted detection of pests, disease and level of nutrition in plants. The AI techniques being developed allow individual weeds to be detected, with just the correct amount of herbicide then being applied with laser precision. This targeted process saves money but also helps to preserve soil quality and stops toxins entering the food system. Some of these techniques are already being used in developed countries, but applying these approaches in Africa will have an enormous impact.

In one example, PlantVillage, a non-governmental organization (NGO) in Kenya that helps farmers with agricultural knowledge and support, has created Nuru, an AI assistant for African farmers. Developed in partnership with the Food and Agriculture Organization of the United Nations, this tool, which runs on any smartphone, uses AI vision-processing to provide expert-level diagnostics of crop diseases. Just by using the camera on their phone, farmers can identify the crop disease and get help on how to address the issue. It also includes an AI chatbot that provides immediate answers to questions posed by farmers. Developments with drones, and from using data captured in satellite images, are also being explored for future applications.

Another key global challenge is fast-rising demand for water. Clean, fresh water is becoming increasingly rare. Today, almost a third of the world's population is living in water-scarce areas.[10] In developing economies, the demand for water is growing fast, and this is being driven by much-needed improvements to sanitation systems but also by the increasing spread of water-consuming appliances, such as washing machines and dishwashers. More intensive agriculture is another major driver.

Fortunately, AI is being used not only to monitor climate and rainfall, but to help us understand moisture evaporation too. The rate at which moisture reduces is termed 'evapotranspiration' – a combination of water evaporating into the atmosphere from the Earth's surface, water being released from groundwater and from lakes and rivers, and transpiration (the movement of water out of the soil through plants and into the atmosphere). The challenge in understanding evapotranspiration is to capture enough real-world measurements from across any given area so that you can build up a realistic measurement. The only solution is to 'interpolate' from the small number of real-world measurements that you have available and fill in the missing information to build up a full surface-water picture. Simulations are helping to solve this problem so that much more accurate predictions can be made, and to understand just how long water reserves will last in specific areas. But by combining these calculations with accurate weather forecasts – also increasingly powered by AI – the moisture conditions can be much more accurately predicted, and irrigation can be brought in before crops become distressed. Looking to the future, this type of AI-driven moisture understanding can also be combined with predictions about climate change to identify areas of land that may become harder to farm, or to identify crops that might better suit the new conditions.

Clearly, much more work on understanding our environment is needed. Creating healthy lives for all must be a priority and artificial intelligence has a role to play, both in addressing the growing conditions and environmental impacts directly but also in helping to create more wealth in other parts of emerging economies. AI can be used to boost productivity and support the development of these emerging economies. This increasing wealth can then be used to solve the fundamental problems once and for all – not only in developed countries, but everywhere.

*

Artificial intelligence is starting to have a very positive effect on our society and is helping to address major societal challenges across education, health and the environment. AI will also improve efficiency in the workplace, taking over dangerous or repetitive tasks and freeing us to focus on more creative activities. Autonomous transport systems are becoming possible, and AI will help us solve many other complex scientific problems that were previously out of reach. But AI is a powerful tool, and we need to understand the challenges that this new technology may bring.

16

THE CHALLENGES OF AI

Eric Schmidt, the former CEO of Google, has said that in the early days of Google he was naive about the power of information. He is now calling for tech companies to become better aligned with the ethics and morals of the people they serve.[1] In early 2023, after the launch of ChatGPT and driven by its sudden impact, an open letter was distributed by the not-for-profit Future of Life Institute, an organization backed by Elon Musk and whose stated aim is the reduction of the existential risks facing humanity. The letter called for: 'all AI labs to immediately pause for at least 6 months the training of AI systems more powerful than GPT-4'.[2] In addition, they said: 'AI developers must work with policymakers to dramatically accelerate development of robust AI governance systems.' Unfortunately, a pause is just not realistic and would just hand a massive advantage to the current leaders in AI systems, creating monopoly outcomes. But we do need to learn from the experience of Eric Schmidt and to understand that, although AI is having a positive impact in many areas, it is also raising many complex questions and must be controlled. However, deciding how to do this will test our fundamental human values.

Your views on ethics, privacy and how you treat others are all very much shaped by your upbringing, your education, your circumstances and place. As historian Tom Holland shows us in his book *Dominion*,[3] despite our increasingly secular society, many people who have grown up in Europe and North America are still very much influenced by

Christian values, even though they may claim that they hold no religious views and never go to church. Over the course of nearly 2,000 years, the values of common decency and tolerance have become engrained. Science, gay rights and even atheism have all emerged as a result of this positive cultural inheritance of tolerance. In Asia and other parts of the world, many of these same values exist but perhaps with some subtle variations influenced by differing histories, religions, local traditions and the equally strong values that have been passed down through generations in these regions. Without even being aware, many are influenced by Abrahamic ideas and others by Confucian principles, and some by other values. As a result, what is right for one place may be different somewhere else and these views may transform over time. It is not for machines to decide what is right; these are tough choices humans must make, and tolerance will be needed. Regulation of AI must work for all, but we need to act fast before AI gets used against us by a malevolent group.

Many social commentators, business leaders, governments and even some scientists claim that AI is set to take over and say this in such a way that suggests the machine cannot be trusted. Although artificial-intelligence technology will bring many new challenges into our lives, we must not abdicate responsibility and blame the machine. We need to look at the emerging issues that are undoubtedly heading our way and understand how we, as humans, can address these. We built the machine and so we need to hold the humans, who developed this new tool, to account. I don't propose to make a definitive set of suggestions here about how we can solve each individual issue; the answers will be complex and the issues will no doubt evolve. We will have to keep working to improve outcomes and learn from experience. I will try to highlight some of the most critical areas, including privacy, bias, playing with our human emotions, the impact on jobs, and weapons. To do this, I will propose an extremely simple framework that can help everyone engage and discuss these issues in a more constructive manner.

FRAMEWORK

Today the technology industry and academia are the primary drivers of progress in AI, while most governments are still playing catch-up. To build trust we will need to establish regulatory frameworks that help to provide appropriate guidelines and controls. Putting these in place will require a collaboration by all these groups but also requires much greater technical understanding within governments so that oversight is not just outsourced to commercial interests. There is an urgent need for an inclusive public dialogue that will help to raise awareness around the issues. To help focus these discussions appropriately, let me propose the following as a simple framework that can help:

1. We first need to understand if the public are *aware* that an issue may exist.

2. Then we need to understand how this issue may *manifest*.

3. Finally, we should look at how people can be *protected*.

I will use this very simple *AMP* framework to review some of the bigger issues that are emerging around AI, first of which is privacy.

PRIVACY

In 2006, UK mathematician and data scientist Clive Humby coined the phrase 'Data is the new oil'.[4] Your personal information is just like oil. It is an extremely valuable commodity that is being used to power some of the largest and most profitable companies on the planet. Unfortunately, we are all giving this very rich source of wealth away for free. We often know very little about how much information we are sharing and who is using it. Did you know, for example, that Google can track your mouse movements and holds a patent for technology that displays paid-for ads based on where you move your mouse even when you don't click?[5] In fact, many companies have a business model that is built around gathering information about us and then using this to power their tech platforms for profit. As the saying goes, 'If you are not paying for it, you're not the customer; you're the product being sold.' Often our personal information is being sold to other companies, to researchers, to computer hackers, to law enforcement agencies and even to foreign governments. How private is your most personal information and how much do you freely share every day?

These days, when you travel to a foreign country you are required to share what is called biometric information. In addition to bio-graphical details (name, place and date of birth, passport number, etc.), many border patrols will collect fingerprints, iris scans, facial images and other biometry. This data contains context, so it qualifies as information – about you. Its primary purpose is to authenticate that you are the person that you say you are. Biometric information has had an extraordinary impact on protecting our borders and on stopping people who were previously able to travel under false names from using false documents. However, you might hope that this highly personal information remains locked up in the border-force offices. Instead, authorities share suspicious biometric information that they harvest at passport control with police forces all over the world, to help keep us safe. We perhaps find this acceptable, but in many countries our information is also shared and used much more widely for other purposes. The country that issued your passport does not control the use of your information – every country that you visit has different laws and regulations relating to what they can do with your personal biometric information.

As another example, during the Covid-19 pandemic we were all asked to perform Covid polymerase chain reaction (PCR) tests that involved taking a swab from our nose and/or throat. This swab not only contains potential genetic information that would signal the presence of the SARS-CoV-2 virus, but it also contains your complete DNA. Your highly personal DNA information is now in the hands of the many private companies that sprung up to deliver these rapid-test services. Did you ever take the time to read the roughly 5,000-word privacy agreement, printed in a tiny font, that outlines what you have agreed these companies can do with your DNA information? Certainly in the UK, the law requires that informed consent must be given before any sensitive medical information can be shared. It is perhaps a question in law if informed consent has really been given through this 'by using this service you agree to . . .' small-print notification.

As you are perhaps also aware, your own personal electronic trail follows you everywhere. Every time you use a credit card or make a purchase on the internet, when you use a search engine, stream music, upload your smartwatch information, use the map on your phone (with GPS information streamed from your phone as you move around), ask Alexa, Siri or ChatGPT a question, or when you post on social media, some company is gathering information about you. In most countries, and by law in Europe, you can search to find what these companies know – and you might be surprised to see exactly how much you have shared.

Information can be used to build knowledge. Companies can discover what you like, who you know, even your state of health. By agreeing to share this information you will receive some valuable services, such as allowing you to find products, services and content that is much more relevant to your interests. You will get information from your friends, allowing you to stay connected across continents and time zones, all for free. Perhaps you will even receive an early warning on some medical test that you need to take or be offered a medical device that could stop you from developing a chronic condition. But this knowledge could also cause you real problems. Spam marketing messages may be annoying but what happens if a health insurance company gets to know that you have a higher risk of contracting a certain critical illness?

Let's try to use our simple framework to discuss issues around AI and privacy.

Q. Are we *aware* that privacy issues may exist and are they potentially made worse by AI?

A. Yes, but much more awareness is required.

Q. Do we understand how these privacy issues may *manifest*?

A. They are often hidden, and many people do not understand the extent to which their information is being used.

Q. How can our privacy be better *protected*?

A. Well . . . let's discuss.

Most people are aware that privacy issues exist around their personal information and that these issues may be made worse by AI systems. Unfortunately, most are not aware of the details of the contract that has been agreed or how widely their information can be shared. More awareness of these privacy issues is required, and we need much greater transparency.

How privacy issues may manifest is also not well understood. Some may be obvious, but many remain well hidden. Did you know, for example, that Mastercard earned over one quarter of its revenue in 2019 from selling services from card-usage information? They claim that your personal data is anonymized, randomized and totally secure. On one level this information is making your card transactions safer than ever, but at the same time all your transactions can now be observed, with marketing companies tracking and analysing purchases in real time. Do you know who is watching and then selling on this information, let alone who the buyer is?

So how can we put in place regulations that warn people of the possible issues and that will provide a level of protection? There are perhaps two key areas to focus on: the collection and sharing of your personal information; and the way in which it is used – especially to build AI systems.

The first step is to require organizations to make you fully aware that your information is being collected and to give you full control over this process. Companies also have a duty of care to protect your information and not to share it without your explicit approval, and they must let you audit the information that they hold so that you can delete it and adjust the methods by which they collect it. This audit process needs to be made simple and transparent. The European General Data Protection Regulation (GDPR) is a good step in this direction, and pushing for something like this to become a global standard may help to mitigate some of the privacy issues. A public warning message (like the health warnings on cigarettes) may be appropriate so that you can be more aware of what you are signing up to and how your information is being used. Standardized processes and freely available tools could also help to provide transparency and would level the playing field for small companies that must compete with much bigger companies that are more able to absorb the extra costs of these privacy regulations.

The second area of focus is around how your personal information is used. Taking a cue from the medical Hippocratic principle *Primum non nocere* (first do no harm), we must first educate the developers of the systems. These people are the most expert and best placed to ensure that the system does no harm with your information. We need to invest in better training, and I would recommend that we include ethics classes in AI and computer science qualifications. National and international medical boards maintain ethical guidelines for health practitioners, and perhaps we need similar new institutions focussed on the tech industry. These should be sponsored by both industry and governments so that they can remain independent. Government regulation must also place a requirement on organizations to follow the principle of 'do no harm'.

As we will learn later, organizations cannot use the excuse that the

AI system is a black box operating outside of their control. Instead, the objectives of an AI system are set by the developers and so companies that take your information in exchange for delivering a service must be held accountable for how they use your personal data. The industry must find ways to become more transparent and take responsibility or companies risk a backlash and even the introduction of more stringent controls. The key to this will be to make the AI systems more human-centric (a topic we will cover in the next chapter). But how can we ensure that our regulators understand enough about the underlying technology so that they can set appropriate rules? How can we ensure that large tech companies don't use their lobbying power to divert regulation so they can maintain their valuable store of information about you? Socially oriented AI institutions that are supported by both industry and government could support government regulators and help provide an environment of trust in which people's rights are better protected.

Data anonymization is a solution that is often put forward as a way to protect your personal information, but we have already learnt that AI is very good at finding patterns. Even when you remove personal identifiers, attackers can still use de-anonymization techniques to trace the original information. Your personal information often passes through many different systems and some of your identifiers can become known. De-anonymization techniques can then cross-reference these sources and use this to reveal information about you that you thought was safe.

For example, your health records may show that you have an extremely rare health condition, and then at some point you move house. The association of just these two facts could easily unpick your anonymized information, and then other personal information about your state of health might become open for others to see. Extra care must be taken, and regulators need to be aware of these subtle technical details. Even though regulation such as GDPR is strict, it still allows companies to collect your anonymized information without consent, use it for any purpose and store it for an indefinite period of time, as long as they remove identifiers from the information.

There are many techniques that can be used for anonymization. These include 'data masking', where some information is altered – for example, by replacing a character with a symbol. It is claimed that reverse engineering or detection becomes impossible with this technique, but not if enough information remains to provide sufficient clues to pick you out through association when your data passes through enough systems. 'Pseudonymization' is another technique, in which your private information is replaced with a pseudonym – for example, replacing your name with a fictitious name. These techniques and others, or combinations of them, improve the security of your personal information – but you still need to know what information people hold, how they are protecting it, how they might use it, and with whom they may share it. Given enough views of these different pieces of anonymized personal information, powerful AI systems could still unpick these privacy tricks.

Facial recognition systems are another potential threat to our privacy. These systems are not bad in and of themselves, even though they can potentially track you as you pass through a public space. It is the rules that are applied around how this information is used, and the trust that we have in the authorities that have access to this information, that is the issue. Are these systems just being used to find malicious people or are they learning about all of us? Perhaps more troublingly in some regimes: when do you cross the line to become a person of interest? This type of facial-recognition information can also be combined with many other sources of personal information, such as who you know, your shopping habits, what you look at on the internet and your travel plans. Government agencies could build a very complete picture of you. In most countries, governments have a right to access information for reasons of national security. The rules for these types of access to personal information vary from country to country. We need to be aware that AI can let authorities learn a great deal about us and as citizens we must hold our governments to account for how they are using our information, just as we must hold large internet companies accountable.

You should also be aware that, depending on which services you are using, your information may be held on computer servers in foreign countries and that these foreign governments may have a legal right to access your personal data. Global consistency here is important and we need to ensure that our personally identifiable information (or PII) – which might include personal data, or facial recognition and biometric information captured at a border post – is not abused. Making sure that nation states can be held to account on these issues will become increasingly important.

The key point is that we cannot rely on the AI system to protect our private information. The machine just learns from the information that is fed to it and AI just follows a method defined by humans. The machine is not responsible for protecting our privacy – humans are. We must hold humans to account.

BIAS

If you were to base your judgement on a historical review of all the Nobel Prize winners, you might easily come to the very wrong conclusion that white males are the smartest humans. As of 2020, over the 119 prior years that Nobel Prizes had been awarded, only 3 per cent of the science awardees were women and none were Black. We live in a biased world and unfortunately humans have historically held strongly biased opinions. Gender, ethnicity, looks, height, the size of your head, the way you dress – none of these things make any difference to your intelligence or your abilities as a human being.

I was at a conference recently and fellow speaker Garry Kasparov, the world-famous chess champion, made the point very strongly that 'the data is just the data – you can't blame the data. The data is not biased, it is the humans who generated the data that are biased.' And he is correct. The real risk that we face is that artificial intelligence will consolidate the historical biases that are already encoded in much of our current information and in the way previous generations wrote their history. There is unfortunately a real risk that we will carry over our historical human biases to artificial-intelligence systems by training them with this biased human information.

Do you know how many decisions have been made about you by AI in the last month? When you send an email, an AI filter will check your message and may decide to push it to Junk. If you send a customer service request, it may be an AI chatbot that is initially dealing with your issue. Whenever you pay with a credit card, an AI system is deciding if you are making a legitimate payment and may block the payment if it decides that fraud is happening. A job application may initially be screened by an AI system, as might a loan application. We must ensure that we are not introducing bias, or we could inadvertently make life very difficult for many citizens.

With artificial intelligence there is also a real risk of undermining cultural differences, of filtering out distinct national characteristics and even losing the diversity of foreign languages. There are many nuances in local cultural references that make us all unique. Our diversity is one of our human strengths and is critical in the evolutionary process. I sometimes find that my ironic sense of British humour does not always land perfectly with people in the USA – 'two countries divided by a common language', as the saying goes. In the UK we use different cultural references and find certain things funny that an American might think are too close to the bone. Sometimes the brain tries to simplify its view of the world, which can lead to a cultural bias against the values or views in other regions.

Psychologists have classified more than 180 types of human 'cognitive bias'. These are typically caused by a shortcut in the human thought process that then affects an individual's judgement or decisions. One example is 'anchoring bias', which is the tendency to rely too heavily on one piece of information when making decisions – perhaps the first or only piece of information that you learnt about a subject. Similarly, you may overestimate the chance of a particular event occurring if this happened to you recently or the issue is emotionally charged. We are perhaps also familiar with 'normalcy bias' (a form of what's called cognitive dissonance), which is a refusal to plan for something that we have never experienced – a global pandemic, for example. Bias also comes from the tendency of less-skilled individuals to overestimate their own ability (for example, social commentators opining on AI) and for experts, who have a better appreciation of how complex a subject is, to underestimate theirs – what's known as the Dunning–Kruger effect[6] (after David Dunning and Justin Kruger, the psychologists who identified it). There is also a 'peak–end rule', where people seem to focus on the best or worst part of an experience and then zoom in on how it ended, but ignore the average experience, which might be much better or much worse – a nice holiday ruined by a bad experience travelling home, for example.

AI developers may unknowingly introduce forms of cognitive bias into the AI learning method, or they may rely on the wrong training information, which can easily introduce these types of effects. Lack of complete information can cause bias by excluding certain groups or sections of the population. As an example, research studies at universities often use the most available and willing source of participants: undergraduate students. Wikipedia contributors might bias towards their own interests. As a result, the study or the contribution may not be representative of the whole population. AI systems developed by predominantly young male engineers may introduce similar biases.

We need to ensure that AI systems do not narrowly focus on just one source of information but instead build up a broad knowledge that includes a wide diversity of inputs. Natural language AI systems will need to be trained on hate speech, otherwise they won't know it when they see it. But defining what constitutes bad language may vary from place to place. Different nations may find different types of speech offensive, and the AI system might also miss the irony in certain language. Saying that someone is 'really clever' may be a genuine compliment or it might be said as an ironic comment and mean the reverse. Nation states will have a role to play to ensure that their language and their culture are well represented through foundational AI systems. They will need to invest in building their own large-scale AI platforms, as a small number of European and Asian countries are already doing, so that they do not become dependent only on the USA or China for these advanced language models.

Bias in AI is a real issue. However, many of the stories that you read today are driven by narrow and incomplete AI systems, as just one frequently cited example shows. In 2016, Microsoft released an AI chatbot called Tay that within sixteen hours of live testing had learned misogynistic and racist behaviours from online information that it was seeing. The system was quickly taken down. This outcome was definitely not driven by bad intent on the part of the developers, it was instead just a weakness in the machine learning method. It was not the machine's fault; it was caused by the developers' human error and by poor testing.

Many of the weaknesses that we might read about come from poorly designed systems. Data analytics systems that use human-programmed deduction, and that too narrowly follow the training information, are particularly susceptible. More complex AI systems can be built that will generalize more and are trained with much larger sets of information, and these systems are able to move beyond the very narrow set of information that is sometimes used for training. As just one example of how these systems are improving, very large AI language models are already becoming better at capturing knowledge about irony and humour, and as a result can understand much more nuance in our human conversations. We should expect more issues to come up as the technology develops.

Learning how we can build trust into these much more powerful AI systems will become very important. Using our simple framework . . .

Q. Are we *aware* that bias may exist?

A. Unfortunately, humans can be blind to the facts.

Q. Do we understand how these privacy issues may *manifest*?

A. It is certainly clear to the people who are affected.

Q. How can we *protect* people from this bias and avoid it happening?

A. More below.

Governments are today starting to focus on AI ethics, but they need to recognize that it is not the AI itself that is biased or culturally insensitive, it is the companies that control the platforms, and how they direct the systems, that can lead to these issues. The methods that these systems use are developed by humans and we need to build additional systems that test the AI. The information used to train these systems is most likely also biased. Regulators need to educate themselves and not be led by ill-informed opinions, but instead follow a more scientific process to identify the real underlying issues. They will need to invest in systems and skills, and recognize that building and using more advanced AI to address these issues may provide the best way to implement new regulations. Governments need access to independent, unbiased advice from new institutions that have the expertise to really understand these complex issues and come up with informed solutions that build trust.

As awareness is raised, a commercial opportunity will develop for AI systems that can unemotionally screen out bias. It's easier to program bias out of a machine than out of a human mind. One approach that has been proposed resembles a consumer blind test comparison where testers are not told and cannot see which product is which so they can't be influenced by their preconceived biases. This type of approach can be applied in AI to assess the impact of adding or removing training information. The results could then be compared by humans to check how different information influences the answers. These results can also be used to look at the learning method to ensure the model is learning appropriately. These tests are not perfect, and if they are poorly designed the results could still be skewed by narrowly focussing people's attention on just one attribute. We will need a diverse test group.

To address bias, it is becoming clear that we need to focus on a few key areas. We must first be aware of our own human biases, whether conscious or unconscious, around issues such as gender and race, but, as we have also seen, there are lots of others of which we need to be mindful. Building systems that can start to test whether these biases have found their way into the AI system will be important. Investing more into research in this whole area of bias should also become a priority for the large internet companies, in particular, whose businesses could be adversely affected by biased systems. Some initial, freely available open-source test systems are starting to appear. As examples, IBM offers AI Fairness 360, an open-source toolkit that examines, reports and helps mitigate discrimination and bias, while AI company Synthesized provides FairLens, an open-source package for data-bias discovery. More companies and entities are starting to offer open-source projects and to contribute to these efforts, but much more work is still needed. We also need AI developers to have a much better appreciation of ethics and values, and we must ensure that the teams themselves are more diverse. Our AI systems must learn from a diverse set of information and be developed by a diverse set of people. Ensuring that the systems are human-centric and also tested by a diverse set of potential users will help, but a fact-based approach must be applied to ensure that this testing is well designed. Making these test systems comprehensive must become a priority.

More education around issues of bias, in particular unconscious bias, is critical. There is a real role for socially oriented AI companies to help in building AI systems that can find bias in training information and that can contribute to open-source initiatives. We need test systems and test information that can identify bias in the outcomes that narrowly focussed objectives may lead to. We need independent institutions that can help to hold companies that build AI-based systems to account. We must ensure that biased information sources do not lead us astray. Using AI effectively could actually make a real difference and might even help to lessen bias over time in wider society as a whole.

HUMAN EMOTIONS

Social media and internet search platforms may know more about you than your friends and family. A study by Cambridge University and Microsoft Research has shown that information gathered from your social media 'likes' can be used to accurately predict a lot about you, including your age, sexual orientation, ethnicity, religious and political views, personality traits, intelligence, your state of happiness, parental separation, and whether you use addictive substances.[7] They might use this information to predict what you like in your content feed, but some systems may also have been trained to play with your emotions. Being aware of how a company might use AI to manipulate you could help you stay in control.

Neuroimaging is starting to shed some light on the complex processing of human emotion in the brain.[8] Emotions are complex, but one structure in the brain consistently associated with them is the amygdala. The amygdala is a multi-layered and multi-modal system that forms a small part of the brain, and part of this system appears to be related to emotions. It is thought that the amygdala evaluates environmental stimuli and drives emotional learning. It also appears to be responsible for helping to store emotional responses in our declarative memory (related to events, time and place, you will recall from Chapter 5) and provides emotional 'colour' to these memories.

The amygdala is at the steering wheel of what we might think of as wireless communication throughout the brain, using chemical messengers. The brain doesn't just use tiny electric signals to communicate, it supplements these with very complex chemical signals that are still not well understood. It has even been hard for researchers to distinguish between true chemical neurotransmitters and chemical activity that is just confined to communication within a cell.[9] Chemical transmission can be used both to encourage (or excite) some functions and to inhibit other activities and suppress certain senses. For example, when you cut yourself, you feel pain but then endorphins rush in to manage the negative signals coming through your nerves. You may also get a dose of norepinephrine (which we more commonly call adrenaline) as a reaction to the cut. This chemical dilates the pupil of the eye, strengthens muscles, speeds up the heartbeat and inhibits digestion, ready for what we call a 'fight or flight' response. Your brain uses at least forty different chemicals, and researchers are regularly finding new candidates.

As an example of the effect that these chemicals can have, in 2006 a team from Loma Linda University in California led research that was published in *Nature* that found that merely anticipating laughter boosted the production of mood-elevating hormones called β-endorphins.[10] As part of this research, one group of people were told that they would watch a funny video, and their levels of stress hormones were monitored and compared with a control group. What really surprised the research team was that adrenaline, which influences stress and is inhibited by endorphins, decreased by 70 per cent even before anything funny was seen. Just the thought of humour makes us feel better. Research from Lynn University, Florida, has shown that laughter also causes a release of the neurotransmitter dopamine, which serves as a reward for the brain and creates a sense of euphoria.[11] Our brain is extremely sensitive to changes and these emotional stimuli can set off a big reaction.

A number of chemicals drive positive emotions. Dopamine's purpose is to motivate your body towards a distant goal. However, it is also addictive, and the effects are fleeting. Oxytocin is released when you have physical contact with others and is related to feelings of love, friendship and trust. Oxytocin has also been found to boost our immune systems, help us solve problems and make us more resistant to the addictive qualities of dopamine. Serotonin is also related to social interaction but is very different, associated with feelings of pride, loyalty and status, while endorphins are released in response to pain and extremes of temperature or stress, and are designed to help you push your body beyond its comfort zone. When you go for a long, challenging run, you release endorphins to help your body cope, and then when you suddenly stop, you can get an endorphin 'high'. This is why exercise can become quite addictive.

One of the reasons that casinos in Las Vegas create lighting systems that make it always feel like it is late afternoon is to confuse your body. They are trying to visually fool you into thinking that it will soon be night-time, which allows you to explain to yourself why you feel a bit tired. But it never actually gets dark in the casino, so you end up staying longer at the gaming tables: even approaching midnight, it still feels like 5 p.m.* The white light that sits behind a laptop LCD screen or in your mobile phone inhibits melatonin production, so looking at your screen late at night may cause sleep problems. The study of our circadian rhythms, which drive our waking and sleeping cycle, is very much in vogue. Getting a good night's sleep is key to maintaining the health of our brain.

Chemical changes can play with the communication pathways between neurons. They add another level of complexity to the information flow that ultimately drives rewards in your body. Just like the casinos, many online tech companies understand how to play

---

* People who do a lot of intercontinental business travel, as I do, will sometimes take melatonin tablets to try to combat the effects of jet lag. I prefer to rely on a combination of rest, exercise and getting some fresh air and sunlight as the more natural way to keep the chemicals in my brain balanced.

with the chemicals in your brain to try to influence your emotional state. The 'ping' on your smartphone that signals a new message will spark a release of dopamine. This gives you a quick boost, but the effect can quickly become addictive, and social media platforms will try to manipulate this to keep your dopamine 'fix' going. Finding ways to engage you as a consumer is a key goal for online services, and to do this they are building tools that understand how to suggest empathy towards you and how to play with your brain chemistry.

Returning to our simple framework:

Q. Are we *aware* that our emotions are being played with?

A. Mostly not, but we will explore some ways this happens.

Q. Do we understand how these issues may *manifest*?

A. An example that we all live with is everyone's increasing addiction to their devices.

Q. How can we *protect* people from this emotional manipulation?

A. Let's discuss.

Today we all rely on our smartphone. When we leave the house, we make sure that we have our device in our pocket or bag. In fact, a team of anthropologists from UCL University in London have identified[12] that people feel that their smartphone is no longer just a device; it has become a place where they live. We shop online, stream content and spend large amounts of time communicating with each other through online social networks. Even when people are physically together, they can easily disappear into their own online worlds.

A machine will never be able to replace a human in a physical one-on-one meeting, but an intelligent machine could become very good at understanding your mental and emotional state and then act appropriately – or inappropriately – on this information. We all send out clues in many different and subtle ways about our own feelings and our state of mind. Body language, eye contact, the tone of language that you use – these all provide 'tells' as to your emotional state. Some people are better at reading these signals than others; a machine could be trained to become an expert.

Electronics and computer science engineer Rosalind W. Picard from the MIT Media Lab used the term 'affective computing' in her 1997 book of the same name.[13] This has subsequently led to the field of artificial emotional intelligence (AEI), and this is now a very active area of research, especially with online companies. It has also led to several fictitious film characters, perhaps most notably the highly intuitive personal AI device Samantha in the 2013 film *Her*.[14]

As we discussed in Chapter 5, today's attention-based transformer AI methods have rapidly driven new breakthroughs in artificial emotional intelligence. These powerful systems can understand sentiment in written text and in spoken language, and are also being used to create systems that build in much more human empathy. This approach is being used to make you feel 'happy' when using online systems and to drive more engagement that will keep you on these platforms for longer. To illustrate, documents came to light in 2021 that showed Facebook took much more notice if you used an emoji as a response to a post rather than just pressing the 'like' button. Facebook CEO Mark Zuckerberg even encouraged users to use the 'angry face' emoji to react to posts that they didn't like, without telling them that this would actually push more of this same type of content at them.[15] Online tech companies continue to search for ways that artificial emotional intelligence can be used to create stickier environments that encourage you to stay longer and encourage you to buy more goods or services.

Remember, it is not the machine that is playing with your emotions, it is the company that created the system that is doing this. We need to ensure that everyone is aware that these online systems may be trying to manipulate us. We can usually recognize this behaviour in another human and so we must become similarly adept at recognizing it in an online system. We should push companies to self-regulate but ultimately, as governments and regulatory bodies become more attuned to this activity, it would certainly be possible to create AI systems that trawl online platforms to look for inappropriate behaviour and then hold companies to account.

In June 2019 the US House of Representatives Intelligence Committee asked Meta (then Facebook), Twitter and Google what their plan was to deal with potential deepfakes during the 2020 presidential election. Deepfake videos are video sequences that have been manipulated – for example, with different words placed into the mouth of a politician or online influencer. As it turned out, deepfakes did not feature in the election – but the risk is real. Manipulation of photographs is almost as old as photography itself, and computer-generated imagery (or CGI) has been used in feature films for decades. The 1977 film *Star Wars* included creations from George Lucas's company Industrial Light & Magic. Photographic and video editing is not new, and the risk of these modified images being used to manipulate or lead people astray has existed for some time. The risk is that generative AI can produce deepfakes much more easily than conventional computer-graphics-generated approaches, and internet platforms are working hard to identify this type of inappropriate content. For example, German artist Boris Eldagsen submitted an image to the 2023 Sony World Photography Awards. Only after his 'photograph' won the creative category did he reveal that the image was generated by AI. AI techniques are going to be the best way to identify these generated images and videos, but it will be challenging.

Large internet companies already use AI systems extensively to search out extreme user-generated content and to perform content filtering. They are highly focussed on building these controls and they are fully aware of how AI can be used to monitor their company behaviour. It is governments that are behind the curve, and we should expect to see much more AI used in regulation across many technology industries to manage the ways in which companies might train machines to manipulate human emotions.

NATION STATES

Both my great- and my great-great-grandfathers on my mother's side were caught up in the Russian Revolution on the side of the White Army. My great-great-grandfather was a politician representing Latvia as part of the Russian Empire. He died in St Petersburg in May 1917, at the start of the Bolshevik Revolution. My great-grandfather was later captured as the Bolshevik Revolution spread to Latvia and was imprisoned and then shot by a firing squad in Riga on 19 March 1919. This family history has engendered in me an understanding that populations and governments must find a way to cooperate and live in harmony or trouble can easily erupt.

Revolutions, and even quite straightforward political changes, typically result from an imbalance that builds up and affects a particular group within the population. As the ancient Greek philosopher Plutarch observed, 'An imbalance between rich and poor is the oldest and most fatal ailment of all republics.' If this disaffected group is led by an effective leadership (often quite a small set of people) and it becomes large enough as a percentage within the overall population, then either through a democratic process, or through some other means, changes in government can happen. Populations want leaders who can bring stability, personal freedom and economic prosperity for all, but to deliver this government leaders must enforce some level of control. Getting the balance right results from good governance.

Cultural norms and our different value systems may also affect the balance that people find acceptable in this state-control-vs-personal-freedom equation. Confucian traditions, which are common in many parts of Asia, tend to favour stronger leadership, and people may accept a higher level of control in return for stability and prosperity. Abrahamic traditions and values in the West tend to value discussion and consultative leadership in a more open democratic system, but still this varies from place to place.

Communication ends up being critical both between nations and

within individual regions and states. Different countries must communicate so that they can share knowledge that will help them fully
understand each other and identify their differences. This will allow
them to find compromises and ensure that more intelligent outcomes
can be achieved when conflicts or disputes do arise. The key is to
build mutual trust, because without trust the risk of conflict increases.
Similarly, inside national boundaries good communication is needed
so that governments can understand the views of the people and steer
governance appropriately. It is clear to me that there was a break-
down in communication between the population and the leadership
of the Russian Empire at the beginning of the twentieth century. The
people in charge weren't listening and this led to less intelligence on
their part – they weren't gathering the information necessary to adapt
and survive. The revolution that affected my forebears was the result.
Interestingly, in the case of Russia, Tsar Alexander II had approved
a set of major reforms that would have started to move the country
away from autocratic rule. He said, 'I do not hide from myself the fact
that it is the first step towards a constitution.' Unfortunately, he was
assassinated on 13 March 1881 by radicals who were frustrated by the
pace of progress, just two days before he had planned to announce his
new reforms. His son and successor Alexander III quickly abandoned
the planned reforms and instead further suppressed civil liberties.
These actions were then continued by Nicholas II, who had witnessed
the assassination of his grandfather first-hand. Nicholas II abdicated in
March 1917 and was killed, together with his family, while being held
under house arrest, by Bolsheviks in the early hours of 17 July 1918.

We should recognize that the free flow of information over the
internet and the emergence of powerful AI could be good for democracy. But it also opens up the opportunity for increased tracking and
monitoring of populations by governments, down to a very granu-
lar level. This could be positive, in allowing governments to build a
much clearer picture of the national mood so that they can learn
where changes are needed and what improvements they need to drive.

Alternatively, AI systems could be used to increase levels of control, to regulate free speech or marginalize sections of the population. We need to be aware that artificial intelligence could be used to improve democratic outcomes but could also, very easily, be used to support authoritarian regimes.

The Universal Declaration of Human Rights from the United Nations that was originally published on 10 December 1948 lists 30 Articles.[16] Article 1 states that 'all human beings are born free and equal in dignity and rights. They are endowed with reason and conscience and should act towards one another in a spirit of brotherhood.' Article 12 states that 'people should have freedom from interference with their privacy respected, within their family, at home, and in their correspondence'. We quickly need to find ways to monitor whether AI is being used by nation states to cross the line on these issues or on any of the other human rights.

The humans that develop AI should take note of Article 1 and ensure that their system acts towards humans in a way that respects their dignity and rights. They also need to take note of Article 12 and make sure that the system does not misuse people's personal information in a way that would violate their privacy. Effective independent AI institutions that help to define appropriate regulations could assist international organizations like the UN in holding governments to account. These AI institutions must have sufficient expertise in AI to keep up with the complex and rapidly evolving issues. Armed with this detailed information, the UN or other similar organizations could use their influence to ensure that governments are not using AI systems to impinge on people's basic rights. Obviously, we also need to hope that the UN and other organizations hold enough sway in these matters and can drive through the necessary changes with national governments.

Many people are *aware* that nation states may use AI for both good and for bad, but the issues that may *manifest* are complex. We will need to help *protect* people from nation states that try to misuse AI. This is going to be hard and will require concerted international action, but leaderships should never underestimate people. As awareness increases of the bad ways in which AI is being used, the mass of the population will start to push for appropriate use and better behaviours. From time to time, populations may need external help. The issues of how AI is currently being used in China is one example that we must explore in more detail. The situation is complex, but I will save that for the final chapter.

Leaders of nation states should take note: history is full of political upheaval. They need to realize that if they use AI to repress or to limit people's rights, groups within the population will still find ways to come together. If these disenfranchised groups become large enough, then governments may fall and revolutions could still happen. Refusing to learn from history just shows a lack of intelligence.

SOCIO-ECONOMIC IMPACT

In the US census of 1950 there were 271 different occupations listed. Only one of those has subsequently been completely replaced by automation – the elevator operator. With the emergence of skyscrapers in the early 1900s, elevators became essential. These early elevators needed skilful control to regulate their speed, and operators were required to stop the lift parallel to the chosen floor, before safely opening the complicated double door mechanism for passengers. Electronic control systems have completely eliminated the need for this job. But research has also shown that if a job is only partly automated, often levels of employment will actually increase. As David Autor, Professor of Economics at MIT, has noted: 'expert commentators tend to overstate the extent of machine substitution for human labor and ignore the strong complementarities between automation and labor that increase productivity, raise earnings and augment demand for labor.'[17] For example, during the Industrial Revolution the introduction of weaving machines drove down the price of cloth, which in turn increased demand. As a result, many more people were hired as weaving-machine operators than had been employed in the trade previously. Unfortunately, the new roles working on the powered weaving machines often went to new workers, not the existing hand weavers, and as a result the wages of traditional hand weavers fell dramatically and their importance became marginalized. Retraining is critical as these transitions occur. But often new technology just improves people's jobs, removing dull and repetitive tasks.

In the late 1970s entrepreneur Dan Bricklin was sitting in a class at Harvard Business School watching a professor draw out a spreadsheet on the blackboard, entering and rubbing out entries, with all the calculations being done laboriously by hand on a calculator. The Apple II computer had just been released and Bricklin realized that this would be the ideal machine for running an electronic spreadsheet with the ability to enter a new number and then have all the calculations automatically update. Together with engineer Bob Frankston, he created VisiCalc, an electronic spreadsheet tool that went on to transform accounting work that had previously taken hours to complete, turning laborious tasks into a single button-push. VisiCalc is widely credited with helping to kickstart the whole personal computer revolution. But rather than killing off accounting jobs, this new tool liberated accountants from long-winded hand calculations, increasing job satisfaction and the contribution that they could make to the business, and as a result the number of accounting jobs has increased significantly. The *Financial Post* reports that there are over 750 million professional users of Microsoft Excel, the world's most popular spreadsheet package.[18] Accountants can now do much more, producing lots of different financial scenarios and testing many 'what if' calculations. This automation tool liberated accountants, turning them from a backroom function into critical members of the business team.

According to a recent McKinsey analysis, almost all of today's jobs have some aspect that could be automated by artificial intelligence, but very few can be *entirely* automated.[19] This suggests that AI may actually increase employment in many areas, in the same way that partial automation has helped to increase productivity and increase job numbers in the past.

AI is very good at recognizing patterns and will perform its task accurately and tirelessly. Similar to the elevator operator job, repetitive tasks that require constant mental attention are the most likely to be completely eliminated. However, if the boring and repetitive parts of a job can be automated, this may actually free people up to focus on the more creative side of their work, increasing their job satisfaction and improving productivity. Software engineering is a prime example. New natural language processing tools are opening up the prospect of having AI generate accurate software code. Creating software will turn into a process of describing the function that you want the software to perform. These new AI-powered tools will greatly increase the productivity of software engineering, making it much easier for anyone to create complex software programs. However, rather than killing off software engineering, AI may greatly expand the number of software developers, just as spreadsheets expanded the number of people performing financial analysis. Instead of just a few people being able to tell a computer what to do, step by step, in a complex software program, AI will allow us all to tell a computer how it can help us solve a problem. Normally unwieldy enterprise resource planning (ERP) tools (those all-encompassing systems that many people are forced to use at work and which combine order entry, inventory management, accounts, expenses and other tasks) could now be much easier for users to control. If built with an AI-powered software development environment, it might become possible for the user to use a simple prompt (just like asking ChatGPT a question) to change the way the ERP system asks for information or how it is displayed. The users will end up back in control of the system, rather than the computer just saying . . . no.

Humans are social, creative and dextrous, and these are all skills that will be hard for an AI system to replace. Hairdressing is unlikely to be replaced by a machine anytime soon. A photographer will use AI to produce much more creative images, but don't expect a machine to replace their role. AI systems will allow much more personalized education, but teachers will still be critical and their job will be enhanced, allowing them to focus on coaching and developing curiosity, creativity and critical thinking in their students. Radiologists will use AI to help review X-ray scans, and the level and speed of analysis will significantly improve, but their job will not be replaced; instead it will change, allowing them to spend more time with the patients, while diagnoses will improve.

AI can drive a very significant increase in productivity and could improve the quality of work across many professions. The internet has had a transformative effect on the way that we live but sharing 280-character messages or pictures of our cats, while fun, doesn't massively change productivity. In contrast, AI can transform work and will drive a very significant increase in global productivity. Our simple framework for looking at the issues that arise can help us to see that *awareness* around the way AI will change employment is still not well understood. On how the issues will *manifest*, we should be much more optimistic and look for ways in which AI can help to improve the quality of work for all.

Some roles, however, could be completely automated. Factory workers, warehouse staff and transportation jobs are perhaps most at risk. Some white-collar analyst roles may also be threatened but in all cases new roles will emerge. We will need to *protect* those people affected by making sure that lifetime education is available that can help people retrain and take on new roles in work that is made better by AI.

The economics of how technology transforms the workplace has been discussed since the time of the pioneering economist Adam Smith with his famous book *An Inquiry into the Nature and Causes of the*

Wealth of Nations, first published in 1776.[20] The experience from the Industrial Revolution suggests that technology transformations drive productivity improvements, which indeed bring significant improvements in wealth for all, but also cause major disruption to individuals and even to whole regions and countries. The more recent experience, following the introduction of computers and the internet, has similarly driven new forms of wealth creation but this wealth has not necessarily benefited everyone. In fact, the effect in Western economies has been to drive more wealth into the hands of the rich, with the people in the middle-income range most squeezed.[21] The owners of capital appear to have benefited much more from this automation of knowledge work than less-skilled or even average-skilled workers. Factors such as globalization have also played an important role in this trend, with roughly 2.4 million jobs being lost in the USA during the period 2001 to 2011 due to the off-shoring of manufacturing that was made much easier by efficient global information networks and the opening up of trade with China (following its entry into the World Trade Organization in 2001). Understanding how AI may affect these trends, and what other factors will become important, is hard to predict. However, it will be important for smaller nations and emerging economies not to fall behind by allowing other countries to gain a major economic advantage from AI. Making sure that training and support is in place for workers affected by automation will be critical, and we also need to find ways to ensure that the benefits of productivity gains are shared broadly. We must find ways to raise income levels for all, and this should include people in developing countries too.

And so, an important key question remains: could AI help to eliminate poverty? AI can perhaps help to create a new social contract. Much better access to quality education will allow more people to learn, and to learn at their own pace. They would then also be able to continue developing their skills throughout their lives. AI will help teachers to have more time to focus on helping people build these critical, lifelong learning skills. AI can perhaps also help to make people more productive and have greater job satisfaction, with fewer boring and repetitive jobs. However, we need to ensure that the benefits from AI technology can be made available to all and that it is not just restricted to rich countries and to large, successful companies. AI can help businesses of all sizes compete more effectively, but this will mean that even small companies and developing countries will need to invest in this artificial-intelligence-driven future.

I am optimistic that AI can have a very positive effect on the way we work; it can increase our creativity (not replace it) and will help us lead more productive working lives. However, the transition may be difficult, and close attention will be needed to ensure that everyone can benefit.

WEAPONS

My grandfather on my father's side was a Quaker and a pacifist, born in London. He was also an engineer. During the First World War, he refused to fight but instead helped to design and build aeroplanes at Martinsyde Limited, a manufacturer of surveillance planes, supporting the war effort as an essential worker. During the Second World War, among many other items he designed a decoy-tank landing craft that was called the Big Bob. Hundreds of Big Bobs were built and used in various activities as part of Operation Bodyguard, a deception plan conceived by the Allied command. The cunning idea was to convince the Axis powers that the landings in Europe during 1944 would happen at Pas-de-Calais rather than at Dunkirk. My grandfather's work ended up having a big impact. Perhaps some of my grandfather's pacifist tendencies have rubbed off on me, as I also think that armed conflict is a poor way to resolve issues. Unfortunately, it does happen, and it is inevitable that AI will be used during conflicts – in fact, it already has.

During the civil war in Libya in March 2020, a 'lethal autonomous weapon' drone was deployed, according to a report on Libya from the UN Security Council's Panel of Experts.[22] The report states that this drone located and 'may' have autonomously attacked and killed members of Libya's armed forces. 'The lethal autonomous weapons systems were programmed to attack targets without requiring data connectivity between the operator and the munition: in effect, a true "fire, forget and find" capability,' the report highlighted. This incident brings into sharp focus the need for action on autonomous weapons, which Human Rights Watch, one of the leading NGOs that focus on human rights issues, has been campaigning for since 2013.

In December 2021 the United Nations Convention on Certain Conventional Weapons (also known as the Inhumane Weapons Convention) again debated the question of banning autonomous weapons. A prior review meeting had happened in 2018 but dodged the question of whether to move to formal negotiations that could lead to a treaty. Unfortunately, a similar outcome resulted from the 2021 meeting, with some countries, which sources say included Russia, India and the United States, expressing doubts about the need for a lethal autonomous weapons (LAWS) treaty.[23] In fact, the US has previously highlighted that there may be certain benefits around the precision that such weapons could provide. The 125 countries taking part in this 2021 UN discussion stopped short of launching work on an international treaty, instead merely agreeing to continue discussions. This feels like a big missed opportunity.

The International Committee of the Red Cross, several NGOs and many leading AI innovators have all been pushing for an international treaty that would establish legally binding rules over machine-operated weapons. Clearly, we are all *aware* that autonomous weapons are possible. Do we yet understand the issues that may *manifest* due to autonomous weapons? The fact that people could be injured or killed, with a government (or bad actor, such as a terrorist) able to claim that they are not responsible because it was the machine that decided, is just wrong. And how can we *protect* people from these autonomous weapons? In this case the answer is simple: as we have done for many other categories of weapons, outlaw them by putting in place a worldwide ban.

I realize it is inevitable that AI will be used to improve weapon systems and to support battlefield situations. Adding AI-powered tracking and identification to weapons will make them more reliable, more accurate, and potentially even safer by limiting collateral damage. However, there are no mitigating issues or benefits that would arise from allowing these AI-powered weapons to autonomously decide on whether to press the kill button or not. It doesn't show much intelligence on the part of humans if we leave the decision over life and death to a machine. We cannot allow humans to abdicate their responsibilities and let them claim that a machine made the decision. The machine will simply be using a method described by humans. It will be relying on information that will not be complete. The decision cannot be fully reductive nor 100 per cent logical, because there will never be enough information to be completely certain. A human must be in the loop, and someone must be held to account for these terrible acts of killing.

There is a clear case for legislation on lethal autonomous weapons and a worldwide treaty is urgently required. It is a tremendous shame that this issue has not yet been resolved. As we will investigate next, we need to build responsible AI.

17

RESPONSIBLE AI

Around 15,000 years ago a hunter returned to the family cave with a wolf and announced to the group, 'I am going to tame this animal so that it can help me hunt.' Family members shouted back, 'It's going to hurt us! It will just eat all our food!' (Maybe they even shouted, 'It's going to take our jobs!') But genetic studies show that today's dogs descend from a now-extinct form of grey wolf from around this period.[1] The reaction of people in the cave is understandable: here is an intelligent animal that can help humans achieve their goals, but it is also wild and will need to be tamed and controlled. The family members were right to ask: *Can we control this new tool?*

Artificial intelligence is not a biological creature but a machine that's developed, built and controlled by humans. The methods by which the machine learns and operates are described by humans, and its actions and purpose can be directed through setting appropriate objective functions. The machine does not have free will; it has no emotions; and it has no survival instinct or hidden agenda (though if you ask a complex predictive text model about these things, its object-ive function is to reply accordingly). AI is a tool whose purpose is fully described by us, which means that it can be directed to do good – but it could also be directed by humans with bad intent or with motiva-tions that are derived from a different set of ethical standards.

There is no question that artificial intelligence is an extremely powerful tool, and we must exercise caution. It is also very complicated and so it may be hard to control, but controlling complex systems is not a new problem. Understanding and controlling machines to make them safe has been happening for hundreds of years. When the first steam trains appeared, people worried that the speed might mean that passengers would be unable to breathe or that they would be shaken unconscious. Speed limits were introduced, but everyone soon realized that the real issue was related to how the trains were driven. The trains themselves weren't dangerous but, without the correct controls, accidents could still occur. We now have trains that can reach speeds of well over 300km/h, enabled not only by advances in the powertrain but by carefully planned and shielded tracks as well as advanced signalling.

The very first aeroplanes that innovators tried to build were extremely unsafe, and the early aviators would regularly fall from the skies, often with fatal consequences. However, bicycle engineers Orville and Wilbur Wright did something different. Starting in 1901, they decided to begin what they described as 'a series of experiments to accurately determine the amount and direction of the pressure produced on curved surfaces when acted upon by winds at the various angles from zero to ninety degrees'. To do this they built a rudimentary wind tunnel at their workshop and used this to test different rudders, propellers and wings. They were trying to understand whether their designs would fly *before* they launched their machine.

> We had taken up aeronautics merely as a sport. We reluctantly entered upon the scientific side of it. But we soon found the work so fascinating that we were drawn into it deeper and deeper. Two testing-machines were built, which we believed would avoid the errors to which . . . others had been subject.[2]

As a result of this intensive work, their Kitty Hawk aeroplane of 1903 was the first powered aircraft to fly, and with it the brothers kicked off the whole scientific field of aeronautics. Orville, reflecting on their wind tunnel experiments, said, 'I believe we possessed . . . more data on cambered surfaces, a hundred times over, than all of our predecessors put together.'

Since these early days, the aviation industry has worked extremely hard to put in place controls and to prioritize safety. Today, airlines provide the very safest form of travel despite jet aeroplanes being some of the most complex systems that we build. The avionics systems in a modern jet airliner are so complex that they cannot be fully managed using conventional control-system techniques. Additional fail-safes and systems that perform extra checks are added. Avionic controls include redundant circuits and systems that check each other's outputs to ensure that all the systems are working correctly. They look for critical issues, and these monitoring systems themselves are double-checked to ensure that they are not giving a false reading. It is now very rare that systems fail, and when they do there is usually a back-up system available. In extreme situations, when an accident does occur, there is a very rigorous accident investigation process that looks at everything to find out what went wrong. Commercial planes all carry a flight recorder, which captures the complete set of flight information. The flight recorder is still called a 'black box' because the original casing of navigational aids was black, though this is now a misnomer as modern black boxes must be painted bright red so they are easy to find at the sight of a crash. Specialist flight-safety inspectors examine all the available information from any incident and then propose changes to processes or to systems, or make additions to pilot training programmes.

The pilots of aeroplanes are constantly monitored both through training and in follow-up retraining procedures. The systems in a modern aeroplane are also constantly checking the pilots as they fly. Modern aircraft use human-in-the-loop systems – with computers checking the actions of the pilots before actuating the flight-control surfaces – so that both the pilots and the flight-control computers are working together to make your journey safe.

Artificial-intelligence systems, too, are sometimes also referred to as 'black box' systems, implying that we can see what goes in and what comes out, but we cannot see what happens in between. We rely on the inputs to generate certain outputs, and there is a suggestion that we cannot explain what is happening in between.

The outputs of a deep learning AI system are probabilistic, just like the outputs that we rely on every day with our own human decision-making processes. This means that occasionally they will give the wrong answer – just as humans do. The difference is that an AI system will not get upset if we put in place an additional system that monitors and checks its outputs. An AI system can also calculate and, if we ask it to, will openly share the probability for its decision in a way that humans are just not able to. We need to get used to thinking of these systems less like oracles and more like weather forecasts – clever, useful, but ultimately probabilistic (yet hopefully more accurate).

The vast majority of issues that we see today in AI systems relate to poor AI system design. An artificial neural network is built to an explicit architectural specification, and it is trained using specific methods. It uses a finite set of training samples and is directed to achieve a well-specified objective. Humans have access to the parameters that the machine learns, so artificial neural networks are not 'black boxes' but are in fact 'glass boxes' that we can look inside. Yes, they are extremely complex and very hard to understand, and yes, they do learn from information rather than being told what to do. However, convolutional neural networks used in vision systems have been analysed and understood, large language models are more complex and perhaps more difficult to analyse, but this is still possible. And it is also possible to build systems that control these complex systems, too.

As the famous physicist Richard Feynman wrote: 'What I cannot create, I do not understand.'[3] We created artificial intelligence and so we should be able to understand it. What we find hard to comprehend is the fact that these machines appear to be learning and operating by themselves. As humans, we associate learning with rational thought and so we assume that AI machines must also be exhibiting some type of purpose-driven thought process. They are, but the purpose is set by the training method that humans have defined and by the task we have assigned. It may be their 'thought', but it remains our purpose.

As we have already seen, humans cannot easily rationalize how we determine the trajectory and bounce of a tennis ball or how we work out the best way to return the shot down the line to win in a game of tennis. If you tried to think about all of this while playing, you would overthink the shot and most likely completely miss the ball.

Humans might try to rationalize what actions they took after the fact, but the reality is that we have evolved and then fine-tuned a highly complex biological neural network with just the right set of parameters, and it is this biological neural network, hidden away inside our brain, that is making a probabilistic inference that then actuates our muscles to make that tennis shot. These human systems are a black box hidden away inside us, but we *can* look inside artificial neural networks to examine them and understand what has been created. And we can also define ways to control them.

In 2014, Jason Miller, an engineer and philosopher who teaches robot ethics at Carleton University in Canada, taking inspiration from ethical dilemmas that are called 'trolley problems' and that are often used in ethics discussions,[4] posed the following and now famous ethics question called the autonomous car 'tunnel problem':[5]

You are travelling along a single-lane mountain road in an autonomous car that is fast approaching a narrow tunnel. Just before entering the tunnel a child errantly runs into the road and trips in the centre of the lane, effectively blocking the entrance to the tunnel. The car is unable to brake in time to avoid a crash. It has but two options: hit and kill the child, or swerve into the wall on either side of the tunnel, thus killing you.

The tunnel problem is posed to us as an insoluble dilemma that forces the AI to choose between two terrible options. This question may be valid for a car driven by a human but should never apply to an AI-controlled autonomous vehicle. The key words in this text are 'fast approaching a narrow tunnel' and 'The car is unable to brake in time to avoid a crash.' We must hold the developers of autonomous cars to a higher standard than this and ask: Why is the autonomous car fast approaching a narrow tunnel at a speed at which it cannot safely stop?

In this situation, it is not the autonomous car that is at fault if an accident happens. I believe that the responsibility clearly lies with the human developers of the AI system. The person travelling in this fully autonomous vehicle has no control and is therefore not liable; they are just a passenger. The liability lies squarely with the company that has developed the machine learning method and who has sold what is essentially a product not fit for purpose – they must be held to account.

As humans, we perhaps believe that we are all 'above-average drivers' and that we are so skilled that we can easily drive our cars fast. In fact, as Mario Andretti, the famous American champion racing driver, said: 'If everything seems under control and you don't come walking back to the pits every once in a while holding a steering wheel, you're just not trying hard enough.' (Though perhaps motor racing is a special case.) However, we must expect more from the developers of these autonomous machines. As a result, most humans are going to find a safe autonomous car quite frustrating. It should be designed to drive with the ability to take account of unexpected situations and to stop safely. This means it will need to travel more slowly than our human emotions might think is appropriate. We have lots of rules and guidelines that human drivers are required to follow but these will need to be modified to produce some that are appropriate for autonomous vehicles. If an accident does happen, we will need to do what the airline industry does and analyse all the factors that caused the incident so that we can make the system safer next time. It doesn't show much intelligence if human developers of these autonomous systems refuse to learn from experience.

If we think carefully, we can find ways to develop appropriate controls. The AI system will not be responsible. Instead, we must place the responsibility squarely with the companies that develop these systems, and we must ensure that appropriate controls and regulations are put in place that will hold these companies to account. The difficult part is that often the AI developers will also be the ones best placed to ensure that their systems are well designed. Like the avionics industry, AI practitioners must recognize the issues that are raised, and they must develop the appropriate processes, systems and controls themselves. The AI industry must be prepared to open themselves up for scrutiny and to hold themselves accountable. As I have already discussed, computer science and AI programs should add courses on ethics so that developers can start to better understand their responsibility. I previously described how medical doctors must take a Hippocratic oath, the principle of which dates back 2,500 years: *Primum non nocere* – first do no harm. Today, ethical codes for medical doctors are regularly updated and issued by national medical associations. AI perhaps needs its own version of these independent institutions, and AI practitioners certainly need a better understanding of the ethics and issues that surround these powerful systems.

The challenge is that regulation is open to interpretation and certain cases may require additional inputs or to set a new precedent. Again, using doctors as an example, in the UK the British Medical Association has a medical ethics committee (the MEC) which constantly debates ethical issues related to the medical profession, the public and the state. They regularly update their guidelines and regulations. Individual hospitals will also have an ethics board who are responsible for making difficult trade-offs that might relate to dealing with constraints when epidemics strike or ensuring that a consistent service is provided. Doctors can refer to these ethical principles, but they are still sometimes faced with difficult decisions that need extra interpretation. To support them, most hospitals use the 'three wise people' approach to provide back-up and support. In particularly challenging cases, three opinions are sought from senior consultants to help inform the decision. For complex ethical decisions of this type, additional support can also come from the hospital's own ethics committee. In AI, the decisions are typically not life or death, but they can still impact on quality of life or cause serious offence. Regulations and guidelines are required, and practitioners need to ensure that these are interpreted appropriately.

Companies that put AI at the centre of their business operations will need to ensure that regulations are followed, and where interpretation is required, a process must exist that will provide the appropriate checks and balances. The challenge comes when the outcome is not clear. The issue of unintended consequences can easily arise where a sequence of steps ultimately leads to a bad outcome. Having a process to monitor outcomes and understanding the chain of events that led to these is important.

There are ways that artificial-intelligence methods can be developed to check the accuracy of the AI systems that we deploy. Around 1790, mathematician Thomas Bayes showed us how we can understand the probability of a set of answers being correct and then work to improve the solution so that you can achieve even better answers.[6] Methods can be developed that assess the steps and processes that the AI is using and can identify parts that are missing. By addressing these missing items, the system can be made more accurate. By looking at the answers, the AI system can produce feedback loops that work to improve its accuracy. This is like the process of learning from experience, where the reinforcement learning agent is working to improve the accuracy of our machine learning model.

So how should we create controls and how do we ensure that our artificial-intelligence systems act responsibly?

MAKE AI MORE HUMAN-AWARE

With responsible AI we are trying to build a system that is amplifying human intelligence. As a result, the AI that is being developed must recognize that a human is in the loop and that the outputs must be human-centric. The answer that an AI system produces will most likely not be 100 per cent accurate and so it may be better to show a list of the most probable answers. This will allow the user to see that the answer is not definite and will allow them to make the final choice. By the machine saying that it is not 100 per cent sure but here are some possible answers, it also helps to highlight the limitations of the AI system and creates a more human-centric approach. Modelling the human interaction of the system, and then testing with a small group, is essential before expanding to a bigger user group. In this testing phase, it is extremely important to make sure that lots of different stakeholders are included, not just a narrow spectrum – testing for bias and for inclusion is critical.

CHECK OUTCOMES USING SEVERAL CRITERIA

A narrowly focussed objective may help to drive a specific business outcome but does not necessarily support a human-centric approach. By using several different criteria to test the AI system, it will be possible to see different experiences and identify a broader set of possible issues. Checking outcomes over time will also be important, and testing with different user cohorts will ensure that the system continues to deliver the right outcomes and is not being led astray. This is especially important if the model is being continuously updated with new user information or is being fine-tuned to improve its capabilities. The testing approach and the range of tests performed should be as large as possible. In contrast, the criteria for an 'appropriate result' must be very precise so that the system can be tightly controlled. If issues are identified after the system is deployed, the root cause must be found and additional tests added so that this issue is tested for in future.

CHECK THE INFORMATION

Machine learning, and the AI systems that machine learning creates, need data to generate the information that is used for training. If the data is insufficient or is lacking context, if the data is biased or if the data is unrepresentative in some other way, then the information that the AI system is learning from may easily lead the system astray. In a human-centric system, it is important to build trust and so sharing the data sources and highlighting any limitations in the system are important. Data can be supplemented, and data can be checked; the key is to ensure that the data is delivering good information that can be trusted. Creating a feedback loop to fully test the quality of information that the data delivers is critical. The information used for training and the information used for testing information are equally important, and it is critical that these two sets of information are different so that we can be sure that the system can generalize correctly. A few samples might confirm that your AI system can recognize a cat on the internet, but if you are trying to recognize a tumour in an X-ray then much greater care is needed. Testing information must stress the system, testing for 'edge' conditions, bias and other potential failures.

Not having enough training information often creates an issue called 'overfitting'. This is where the model is unable to generalize correctly because it has not seen enough training samples to build a general understanding. If your vision system has been trained only on a limited number of cat types or perhaps only on images where the lighting is completely perfect, the system may only be able to recognize those cats that it has actually been trained on, not any type of cat or a cat that is only partially visible. One famous example was an AI system that was trained to recognize wolves, but it turned out that all the training information used showed a wolf against a snowy background – what the AI system learnt was to recognize any animal standing against a snowy background as a wolf. The system may appear to work in certain very narrow cases but will often fail when real-world information is used. Ultimately, the training data needs to deliver sufficient information so that the model can be trained to cope with all of these edge conditions. It is also incredibly important to understand what limitations the AI system has, either as a result of the information that the contextual data provides or as a result of the training method itself. Users need to be made aware of these limitations and the system must only be used for the intended purpose. Badly designed and badly tested AI systems will over-promise and under-deliver.

TEST AND THEN CONTINUOUSLY MONITOR

Just as in avionics systems, it is usually possible to design a robust test infrastructure that also includes a system to continuously monitor the AI system once deployed. New AI approaches are being developed that can test and check these deployed systems. This is an important and growing area of new research. Testing systems must try to predict, prevent and detect errors before they happen, and certainly before they negatively affect human users. Errors either need to be blocked or a fail-safe must be quickly triggered. Self-checking AI will become an even more important area in the next few years as AI systems become ever more complex.

It is not just AI developers that need to look at how systems can be improved; everyone has a responsibility. Often, we try to ignore technology because it is complicated. In the case of AI, this is not acceptable, and we all need to be aware when we are using AI systems so that we can try to educate ourselves about how they work, what their limitations are, how to use them for the appropriate task, and how the AI system could be improved. We have a voice as consumers of these systems to demand that they don't just try to grab our attention, but provide a positive, human-centric experience.

There are many ways in which AI is already being used for good and it has the potential to help us solve some of society's fundamental challenges. AI can augment our human intelligence. But it is also a very powerful tool, and we must put in place appropriate controls.

We are at the very start of a revolution and, with the right guardrails in place, AI is going to have an enormously positive impact on our lives.

18

HOW TECHNOLOGY REVOLUTIONS HAPPEN

On the evening of 26 December 1878, a crowd gathered in Philadelphia outside the Wanamaker department store to watch as electric lights illuminated a shopfront and a store interior for the very first time.[1] The small crowd watching was able to catch a glimpse of the electric future that lay ahead. We think of revolutions as being sudden, like a light bulb pinging to life, but technological revolutions actually tend to creep quite slowly into our lives.

We could point to Thomas Newcomen's atmospheric steam engine of 1712[2] as the start of the Industrial Revolution, but it took over 150 years for this revolution to completely change our world. Newcomen's machine was extremely inefficient, and the only suitable application was pumping water from coal mines, where the enormous amounts of coal required for these early steam engines were readily available. It wasn't until 1769, when James Watt delivered his significantly improved condensing boiler technology,[3] that steam engines started to become more generally useful. Watt's partner and financier Matthew Boulton came up with the brilliant idea of charging coal mines a percentage of the savings in coal that their new steam engine could achieve, and they quickly displaced the earlier atmospheric machines. But it was also the smoother operation of Watt's engine that allowed it to be connected to a drive shaft to deliver rotary power to wheels, which led to the more general use of steam power in industry. It still wasn't until 1804 that Richard Trevithick demonstrated his self-propelled railway steam engine,[4] and until 15 September 1830 that the Liverpool and Manchester Railway opened as the world's first intercity railway.[5]

Michael Faraday, working in the basement laboratory of the Royal Institution in London,[6] created the first electricity transformer and the first electric dynamo in 1831. But it then took until 1882 for the dynamo and the steam engine to be combined in the first commercial Edison electric-generating stations, which appeared in London during that year.[7]

Just as steam engines were first used as what we now call 'point solutions' (in this case pumping water from mines), the same was true for electricity. The first application was powering filament light bulbs to replace dangerous and smelly gas lighting. But by the turn of the century, in 1900, electric lighting was still only used in around 8 per cent of urban American homes. Electric motors represented less than 5 per cent of the country's mechanical drive capacity in factories and it still took another two decades for electrification to reach 50 per cent market penetration, with many homes in America not getting electricity until 1945 (by which time penetration had reached 85 per cent).

It is also very easy for us to take today's advanced technology for granted. Every morning we roll over in bed and check for any new messages that have arrived on our devices and click an app to see what happened in the news. But if we still had to rely on electronic valves, the technology that is packed into your smartphone wouldn't just fill your bedroom, it would fill a data centre the size of England, covering an area of around 130,000 square kilometres. There are over 4 billion smartphones in the world,[8] so the data centre needed to accommodate all these devices would cover the whole surface of the Earth, 1 million times over. This unbelievable reduction in size, and the amazing improvements that have resulted, has been achieved in the space of just one lifetime. These incredible devices, which are now part of our everyday lives, wouldn't exist without the advanced semiconductors that power them, without software, and without information theory.

If we date the start of this information revolution to around the invention of the computer in 1944 and the transistor in 1947, then, using our 150-year Industrial Revolution timeline as a guide, this suggests that we are still in the very early phases. The integrated circuit of 1960, and the personal computers that were created as a result in the mid 1970s, are perhaps analogous to the breakthroughs of Watt and Trevithick. As an example, the opening up of the World Wide Web in 1993 resulted in the sale of the first book on Amazon, in July 1995 (a technical book called *Fluid Concepts and Creative Analogies: Computer Models of the Fundamental Mechanisms of Thought* by Douglas Hofstadter[9]), which is perhaps similar to the birth of intercity railways, which also changed the way people live and work.

Sixty-five years after the transistor, the first deep learning artificial-intelligence breakthroughs occurred. Let's align this event with the birth of electricity generation, which kicked off the second wave of the prior Industrial Revolution. Using this marker, our current information revolution is running roughly twice as fast as this former Industrial Revolution. However, the progress from Faraday's breakthroughs to the broad adoption of electricity as a general technology took roughly ten decades, and so, maintaining our 2x rate of innovation, we might expect AI to be a completely general-purpose technology that drives our everyday lives by around 2050.

It is clear to me that, however revolutionary AI appears today – and there are good reasons to think it is – we are still in the very early phases. With today's AI, we are performing the equivalent task of replacing gas lights with electric light bulbs. AI has not yet fully transformed industries. Industries and businesses are still applying AI as point solutions to improve their current processes. ChatGPT is starting to turn search engines into chat engines, but the full potential of this AI technology has not yet been realized.

This slow rate of transformation is perhaps also the reason that we have not yet seen major productivity gains from the information revolution, other than for a short period during the 1990s. As the Nobel

Prize-winning economist Robert Solow has famously said, in what is now known as Solow's Paradox, 'we see computers everywhere but in the productivity statistics'.[10] The general-purpose technology transformations are still to come.

But just as the crowd watching Wanamaker's store light up gained an insight into the future industrialized world that we now live in, so the much larger crowd who tried out ChatGPT in early 2023 were able to see our AI future.

WHAT WILL THE AI-POWERED INFORMATION REVOLUTION LOOK LIKE?

Just as James Watt could not have known the precise impact that his improved steam engine would have, nor could Michael Faraday have imagined the effect of his research into electro-magnetism, so today it is hard to accurately predict the changes that AI will bring. I have already given some examples in energy, in education and in health, but let's examine how AI may end up changing our technology landscape. To do this, let's look at the way you might end up programming a computer with the help of AI.

According to a recent 'State of the Developer Nation' report, there were 24.3 million software developers in the world in 2021.[11] This represents less than 0.7 per cent of the total global working population. Software programmers are a bit like the monks who worked in monasteries, hand-crafting books during the Dark Ages: they represent a tiny, highly educated set of experts who are responsible for highly skilled, laborious, frustrating and often repetitive work. Just as paper and the movable-type printing press changed the way we create books and share information, so artificial intelligence has the potential to make it possible for many more people to program a computer so that it can do useful work for them.

ChatGPT is starting to give some insights into how this may

happen, but a fundamental change will not result from natural language models generating computer code that only helps today's expert software programmers. That would be just another point solution. Instead, what we should expect to see is a much more fundamental systems change. The way we tell computers how they can help us solve complex problems will be radically different in the future. Let me provide a simple example that takes us back to my description of writing my version of the Space Invaders program on my first computer.

Generative-image AI models can already create pictures from a simple text prompt. Generative-language systems can produce text or computer code. The challenge is how to focus the system on a specific problem and on creating a program for a specific computing platform so that it can generate useful and complete solutions.

Very soon we will have tools that allow anyone to create a simple game for their smartphone. These tools will require just a few text prompts – a little like writing a short story – to create your own game. The tool will provide a framework that guides us through a few simple steps to ask what characters we want in our game – using AI to generate a selection for us to pick from. This could be like the conveyer belt in a sushi restaurant, where you just select the characters that you prefer. A few more prompts will allow an AI model to generate the backgrounds and the game environment. A little more guided storytelling will allow the AI system to offer up a set of game-playing scenarios with different levels of difficulty. As you try out the resulting game, it will be easy to go back and change characters, backgrounds and game scenarios. Generative AI will replace the laborious work of creating the software subroutines and will stitch these together in creative ways, guided by you as the developer. Once created, our joint human–AI-generated game can be shared, and others could make suggestions for changes or produce spin-off games. AI will make it easy to introduce these variations and people could easily work together to make games more exciting and more complex.

Rather than having just a few clever people locked away inside software companies, writing complex software code that only a few can understand, AI will provide a new level of abstraction that allows anyone who is interested to program a computer. Artists and designers will be able to take control and try out different ideas. People who enjoy playing games will be able to change the parts that frustrate them, or that they find too easy. By directly connecting the creation prompts to the visual game-playing experience, the creative process will become much more direct. The development process will be like an artist drawing with a pencil on paper, able to immediately see the result of each new pencil stroke.

Games are just one simple example that we can easily visualize, but the same approach could apply to changing the enterprise resource planning software at work (as we already discussed) so that it can become much easier to use and can provide the detailed information that each user needs. Lawyers will be able to create resources that help them manage their case workload and point to legal precedents. As an example, a busy lawyer will be able to dictate a few high-level ideas and messages from which the AI assistant will produce a client letter and an outline brief. Hospital managers will be able to create systems that allow them to manage treatment schedules and predict workloads for doctors. Scientists will use AI to explore exactly how molecules and cells work. Instead of becoming software engineers, they can stay focussed on being experts in chemistry and biology. The encoding required to talk to and control a computer will turn from complex binary computer codes and specialist software programming languages into natural language descriptions guided by frameworks that direct you to create the right type of prompts so that even you will be able to create useful tools.

AI-framework-generated software code will just be a continuation of what we have already seen in software's evolution, moving from complex machine-level logic descriptions, to Grace Hopper's first program languages, to object-oriented programming approaches and on to the way that most software developers today just stitch together higher-layer functions. As each step in software's development opened up the art to new developers, so AI will make it possible for anyone to tell a computer what they want it to do.

GENERATING ECONOMIC VALUE FROM AI

In economic theory, institutions are required to ensure that the economies of nation states can work effectively. Some level of control is needed to establish the trust that is required for economic exchange to happen. This is especially true during times of technology revolutions. Owners of wealth are typically not prepared to risk their money if the investment contract they are being asked to sign has no basis in law or could be terminated by a nation state that keeps changing its regulations. Uncertainty and a lack of mutual trust holds back economic growth.

Equally, if investments into new businesses and new innovations can easily be squashed by larger, monopoly players, then investment and innovation are also held back. If governments rely too much on the existing large commercial companies to create the regulatory environment, then public good might lose out to commercial interests. Commercial companies may then use regulations and lobbying to help maintain their monopoly and make it difficult for new entrants.

People and governments also need to trust that the use of information is well regulated. If AI systems are developed that do not act appropriately, then the exchange of information that leads to knowledge may be held back. For the AI industry to move forward at pace, we will need to build rules that keep our information safe and ensure we end up with responsible AI.

Ultimately, we'll need new institutions to balance safety, consumer interests and innovation. If these institutions are run correctly and are steeped in enough knowledge, then they will actually encourage more investment and drive more innovation. Creating this type of environment will also deliver more investment for both existing and new companies, which in turn will drive more competition and creativity, pushing the industry forward. Investment must not be restricted to established companies, nor to large economies, nor to solving only First World problems. Foundational technologies should be shared so that some of our big global challenges are addressed.

The aeroplane industry recognized early on that independent regulation was critical – passengers' lives were at risk. If a social media company mislabels a type of cat, no one dies (at least, let's hope not). But we must look ahead and realize that AI is an extremely powerful tool. Just relying on commercial companies to set their own rules will not end well. Governments that don't understand the technology will also fail to put in place appropriate controls. For AI to thrive and achieve its potential of augmenting our intelligence, we will need trust in the system.

The European Union is already working on a proposed law that tries to address risk around artificial-intelligence systems. This will possibly become the first legal framework on AI from a major governmental body and it defines three key sets of risk. The first is around systems that may impinge on human rights. It defines these as representing an unacceptable risk. Systems that fall under this definition would be banned. The second relates to potential high-risk applications, such as scanning people's job applications or scrutinizing loan requests. These systems would become subject to specific legal requirements including proposals around 'explainability'. This may end up being problematic. Explainability requirements could drive companies to develop deductive systems that, perversely, make the AI systems less robust; they might produce answers that are wrong, and the proposed laws could end up restricting innovation.

The third set of systems that fall outside these two defined categories are left unregulated under these proposed EU laws. At the time of writing, the specific categorization of the risks and what systems may fall under their remit are still not clear. Some have suggested a flexible system where we can reassign the list of restricted or high-risk systems over time. All of this requires considered and knowledgeable input.

It is in the interest of industry to support the establishment of new, socially responsible AI institutions, with governments also engaged, so that the correct level of independence and trust can be established.

*

Progress in artificial intelligence is racing forward at a rapid pace – every few months we see new breakthroughs – but we are still at the very beginning of the revolution. If used correctly, AI enhances our own human intelligence. AI is not an independent conscious mind; it just uses methods that are described by humans. That means it is under our control, and when things go wrong, we don't have the right to blame the machine. It is the humans who develop and deploy these systems who are responsible.

AI is perhaps the most useful tool that humans have ever built, helping us solve incredibly challenging problems that are currently out of reach. If we use it well, it will reshape how we live and make a better world, one with abundant net-zero-carbon energy, a revolution in healthcare, personalized education for all and greater productivity. AI will support so many critical applications that can improve our lives. But we must also be careful. AI systems, and the companies and governments that develop them, need to be kept under our control so that this amazing technology can become a force for good.

19

MAKING AI WORK FOR US

Stephen Hawking, the renowned British physicist, said of artificial intelligence: 'It will either be the best thing that's ever happened to us, or it will be the worst thing. If we're not careful, it very well may be the last thing.'[1] Using more colourful language, Elon Musk, the controversial entrepreneur, has said we need to be very careful with AI, suggesting it is potentially more dangerous than nuclear bombs, and that adopting AI is like 'summoning the demon'.[2] Many other commentators have made similar statements, warning of the risks that AI may bring.

Many of these comments appear to derive from a highly influential 2014 book called *Superintelligence*,[3] written by the Oxford University philosopher Nick Bostrom. This well-written and important book explores, at a philosophical level, the existential risks posed by extremely powerful artificial-intelligence systems. Bostrom correctly points out that we must anticipate issues and develop ways in which we can control AI ahead of time. However, to make his case, Bostrom also relies on certain assumptions about how AI will develop, including that:

- Computing power will continue to increase exponentially for the foreseeable future
- AI systems will become capable of programming themselves and will be able to independently follow high-level objectives (essentially, that they will be 'sentient')

- AI systems will have the ability to control their own physical environments, either directly or through manipulation

These broad assumptions are not correct, and in reading this book I hope that you have learnt that AI will not suddenly 'wake up' and outwit its masters.

We have seen how computing systems evolved over the last sixty years, driven by the massive improvements in semiconductors and software. I have highlighted how deep learning systems first came to prominence around 2012 and that over the next decade AI inno-vators have been able to make incredible progress, with much more still to come. But progress is not infinite, especially in semiconductors, and we cannot expect the same historical rates of improvement in computers going forward, at least while new computing technologies, including quantum and molecular, are still emerging.

Artificial intelligence is driven by a method developed by humans, and so even when these new computing approaches do start to become viable, AI will still be directed by us. Reinforcement learn-ing and other probabilistic approaches do offer the potential for AI systems to learn from their environment and to improve them-selves, but only in relatively narrow domains where the environment can be understood. The objectives will still be set by us. There is also no easy way for AI to physically control its environment unless we add robotic arms or other physical systems, but again the func-tion and purpose of these would be defined by humans. There are examples of AI systems asking humans to answer 'CAPTCHA tests' (Completely Automated Public Turing test to tell Computers and Humans Apart), which are those picture-box questions sometimes posed on websites to check you are not a robot. But these examples were created by humans, building them into the AI system. The bottom line is that AI is a tool – a very powerful tool – that has been created by humans and can be controlled by us. Like many important tools, it can help us survive and prosper as a species, but placed in the wrong hands or driven by the wrong people it has the potential to do harm.

There are still many limitations in AI systems, and although breakthroughs are happening all the time, technology limits will ultimately also help to keep AI in check. Of course, we must take note of the parallels in history, as Stuart Russell, the University of California, Berkeley, professor and noted AI authority has highlighted:

Some have argued that there is no conceivable risk to humanity [from AI] for centuries to come, perhaps forgetting that the interval of time between Rutherford's confident assertion that atomic energy would never be feasibly extracted and Szilárd's invention of the neutron-induced nuclear chain reaction was less than twenty-four hours.[4]

However, AI is not a physical process that we are trying to understand but a method that allows a machine to learn from information. AI still needs huge amounts of information and massive amounts of compute, and although methods will be more efficient and more automated over time, we will still need to understand how these methods work. One leading researcher who takes this view is Meta's chief AI scientist, Yann LeCun, a leading AI academic and authority, and a Turing Award winner. He argues that we shouldn't fear AI, but instead have people working hard on AI safety, saying in 2023: 'Will AI take over the world? No, this is a projection of human nature on machines.'[5]

There are, of course, substantial risks that stem from AI and so we must be cautious, by putting in place appropriate controls. But I also believe that perhaps the biggest near-term existential risk posed by artificial intelligence is that our leading commercial organizations and our democratic nation states might fail to innovate and could rapidly fall behind in the next wave of AI technologies. If Western governments set regulations that limit AI innovation, if we fail to invest as we should, then we could easily run the risk of other nations gaining a significant advantage. This would not just impact our economic competitiveness but could also have major implications for national security.

As an example, large language models provide a unique ability to quickly analyse huge amounts of text and language. National security agencies are starting to use this technology to scan material to find bad actors and identify hidden security threats. It is easy to see how, without access to the most advanced language AI, our national security could quickly become compromised. Other security agencies, or terror organizations, could generate disinformation or gain unique insights that would put them at an advantage. Even if we work closely with our allies, if we end up depending on them for this technology we could be at a disadvantage. They could delay access to the next-generation models to gain an advantage over us. Maintaining leadership in these critical technologies will become extremely important for the sovereignty of individual countries.

Vladimir Putin has stated that 'whoever reaches a breakthrough in developing artificial intelligence will come to dominate the world'.[6] I find it strange to be agreeing with Putin on anything, but in this case I tend to agree (on the proviso that we change 'come to dominate the world' to 'gain a massive strategic advantage').

We should also recognize that tied up in China's current ambitions to grow economically and build its global influence is a realization that at a certain point in history it fell behind. It is easy to forget that for at least 1,000 years China had the largest economy in the world and was arguably the most innovative nation on Earth. China developed an information economy hundreds of years before anyone else. It invented paper and the use of movable type in printing long before Europe or any other country. Its artworks, theatre and architecture were a match for anyone, and its scientific achievements were of the highest level. China invented the compass, gunpowder, bronze and iron casting, porcelain, the mouldboard iron plough, the horse and ox collar, and seed drills, along with many other breakthroughs. It held a technology lead over the rest of the world for many centuries.

However, towards the end of the Qing dynasty, in the second half of the 1800s, China started to fall behind as the Industrial Revolution ramped up, first in the UK, then in Europe and followed by the USA. China's technology had not kept up, and in various conflicts it lost territories and authority to Britain, France, Russia and Japan. The Chinese refer to this period – which started with the Opium Wars of 1839–1842 and ran through to the end of the Second World War and its own civil war in 1949 – as their 'Century of Humiliation'. It is held up by many in China as an example of why they must innovate and increase their global influence, and why they should never again allow themselves to fall behind in technology.

With this as a backdrop, artificial intelligence has taken on a special significance in China. Kai-Fu Lee, a world-renowned expert in AI and now an investor in AI businesses across China, highlighted in his book *AI Superpowers*[7] that China experienced an AI 'Sputnik moment' in 2016.* DeepMind's success at Go, first beating the Korean champion Lee Sedol and then defeating the Chinese prodigy Ke Jie, raised broad national awareness in China of the incredible potential of AI. Although barely noticed in the US and in Europe outside the tech community, 280 million Chinese viewers tuned in to the televised five-game match against Lee Sedol and then watched as their own Chinese champion was also decisively beaten by AlphaGo. The win sparked a major reaction from Chinese venture capital investors such

---

* A 'Sputnik moment' refers to the national surprise of the USA on 4 October 1957 when the Soviet Union announced the successful launch of the world's first Earth-orbiting satellite, Sputnik 1. Up until that moment, the US believed that they were far ahead in developing rockets and space technology. The announcement of Sputnik sent the whole nation into shock. It was this singular event that led to the formation in 1958 of the National Aeronautics and Space Administration (NASA).

as Lee, who started pouring money into Chinese AI companies, but it also gained the attention of the ruling Chinese Communist Party, and of President Xi Jinping.

In July 2017, roughly one year after this defining AlphaGo event, China's State Council released the country's strategy for artificial intelligence, titled the 'New Generation Artificial Intelligence Development Plan'.[8] This document sets out China's ambition to become the world leader in AI technology by 2030 and to create a trillion-yuan AI industry in China (equivalent to approximately $150 billion). The plan is also closely linked to China's 'Made in China 2025' initiative, which is a national strategic plan and industrial policy aimed at moving China away from being the 'world's factory' towards becoming a technology powerhouse. The use of artificial intelligence is described in the document as the country's 'main driving force for industrial upgrading and economic transformation'.

Perhaps surprisingly for some in the West, China's AI Development Plan also describes how it intends to become the global driving force in defining ethical norms and standards for AI. It sets out principles for the governance of AI that include enhancing the well-being of humanity, respect for human rights, privacy and fairness. China highlights the importance of transparency and the need to be agile and adapt to emerging AI threats. A key difference in these regulations is the emphasis on collective impact rather than just individual rights. As one example, China is leading in facial recognition systems, which are being used for safety and security in China's smart-city projects. Other nations are more circumspect on this technology because of its impact on individual personal privacy, and due to this different approach Chinese companies now have a clear lead in this field of AI.

Although it is a Chinese central government plan, the actual innovation and transformation in AI is expected to be driven by commercial organizations, with strong support coming from local government. Timelined objectives helped to drive focus, and after just one year Chinese venture capital funds had responded, making China the most active country for investment in new applied AI companies, nearly matching the rest of the world combined on AI venture capital investments. The Chinese government have also selected national AI-champion companies to lead in key sectors. Although these national champions gain an advantage, competition is still encouraged and this has driven substantial investment into areas such as autonomous vehicles, smart cities, computer vision for security and for medical diagnosis, and in AI for education. Start-up companies can receive support and subsidies from local government, and you see many Chinese start-up companies building up offices in multiple locations to benefit from regional grants and to work around the restrictions on talent moving from city to city, imposed by their 'Hukou' permanent-resident-registration system.

China's AI Development Plan is not just a major strategy for enhancing economic competitiveness but also to protect national security, as well as for internal security and policing. One key issue that China is having to deal with is that its industrial labour force is no longer growing, with the shift from agriculture to manufacturing reaching maturity. China's citizens have benefited from rapidly increasing prosperity, and a shift to an innovation society is seen as crucial for maintaining this high level of growth. A slowdown would likely make it harder to maintain the high levels of support currently enjoyed by the Chinese Communist Party. To reinforce this point, a pre-pandemic independent global Ipsos-MORI poll showed that, when asked 'if their country was going in the right direction', 92 per cent of the Chinese people who were polled agreed – double the world average on this question.[9]

Chinese citizens have experienced huge structural changes over the last forty years, and I have seen this myself, first-hand, through my regular visits there over the last three decades. In addition to new highways, world-class airports, high-speed trains and the rapid development of major cities, perhaps one of the most significant changes has been the recent emergence of what in China is called 'social governance'.

In 2018 China introduced the Personal Information Security Specification, which provides strong protection for personal data and empowers citizens to control their own information. On the surface this specification appears to go even further than the EU General Data Protection Regulation (GDPR). However, a closer look reveals that exceptions are possible that allow the government, and certain Chinese companies explicitly endorsed by the government, to work around these restrictions. These loopholes allow the Chinese government to exercise very tight control over commercial organizations in how they operate, ensuring that they do not misuse citizens' data but also allowing information to be shared with the government when deemed necessary. This approach is very different from the much freer rein that companies in the USA operate under, where we are asked to trust these large commercial organizations; in China, citizens are asked to trust the government, with loopholes that provide sweeping powers for mass surveillance.

In China, citizens use their mobile phones for everything. It is not just a communication device or an internet connection, but also people's wallet and ID card. Even when you try to buy something from a street trader in a major city, they will only accept mobile payments, not cash nor even credit cards. Personal information can flow from the 'approved' internet company's platform to and from the government. As an example, many vending machines in China include a camera that will automatically recognize your face and bill your mobile wallet without you having to enter any details. Your picture for this AI-powered facial recognition comes from a central government database, and your transaction may also be logged by the government. As a foreigner entering China, I must have my photograph taken by a government machine before I can go through immigration. What you buy, where you go, and what you say in public forums could be tracked. The Great Firewall makes foreign internet websites unavailable, so you are forced to use Chinese sites. The large digital footprint that citizens leave in their wake is also being linked to a new social scoring system that is starting to be introduced. This social governance system is not necessarily seen as a bad thing by many of my Chinese friends, who call it a 'social trust' system and see the benefits it delivers in convenience and for the eradication of crime.

However, it is also clear that AI is being used across China for internal security and policing. Xinjiang province is one region where incredibly tight controls have been put in place over the local Uyghur population. Clear and problematic evidence of excessive regulations and human rights violations is coming to light. The United Nations recently published an assessment that identified several clear issues.[10] What's happening in Xinjiang province is clearly very wrong, fails the test of intelligence and is undermining the trust that other nations have in China.

Artificial-intelligence technology is at the centre of this social governance approach. As a person who comes from a democracy where individual privacy and freedom of expression are key cultural values, I find it hard to see the positives. Looked at through the lens of a law-abiding middle-class Chinese citizen who has grown up with Confucian values, there is perhaps a different perspective: the Chinese see safe cities, law-abiding populations and world-leading economic growth. In the West, we would probably push back against this type of government-driven citizen-scoring system, but we should recognize that our regulations allow commercial internet companies to hold similar levels of information on each of us and leave the power in their hands. The challenge in China is: when do you cross the line and end up having your ability to post stories online, travel, or even just buy goods at the shop, turned off? The issues for Uyghur citizens and other Muslim minority groups in Xinjiang is at a completely different level.

One thing this approach does create is a very different regulatory environment for many AI development activities, especially for those that align with government priorities. China will try to use this alternative approach to secure a technological lead. What we must ensure is that we do not get too tied up with inappropriate regulation and as a result end up falling behind. Just as the Industrial Revolution produced winners and losers, the AI revolution could well cause big differences to emerge between the countries that do develop leading-edge technology, apply it to drive their economic growth and use it to secure a strategic advantage, and those that don't.

Recognizing that AI is complex and requires careful control is correct. Thinking through the potential global impacts is also important: we must ensure that we build AI that is human-centric and that supports our cultural values. But we also need to implement appropriate regulation that does not become overshadowed by scaremongering claims that AI will rapidly run out of control. We need to recognize that other nation states are working hard to innovate and our country must not fall behind.

Today, commercial companies in the USA are driving AI development at a rapid pace. In China, AI is seen as central to its plans for future prosperity and global influence. The UK, Europe, and other regions, cannot afford to take a back seat. They need to invest, too.

AMPLIFYING OUR HUMAN INTELLIGENCE

Over the course of roughly 3 billion years, evolution allowed a single-celled microbe to divide and evolve, eventually leading to the emergence of Homo sapiens and the enormous diversity of life on our planet. Evolution uses a reinforcement learning process that introduces small mutations at each evolutionary step and then 'learns' which works best through a process of natural selection. DNA communicates this learning down the generations, but Homo sapiens also communicates through other means (just as many other species do in their own way), by passing information and knowledge from one generation to the next through language and education, as well as through tools that are taken up and improved by our children.

We collect data about the world around us through our senses. We organize the data into information, then convert information into actions that allow us to adapt to our environment. Ultimately the information that we learn, and the way we adapt, allows us to build knowledge that can help us survive. Our purpose, as biological creatures, drives us to collect information so that we can understand more about our environment and survive. All biological life uses energy to convert information into knowledge, then uses this knowledge to find food, which can provide more energy. Information helps to capture knowledge, which builds intelligence, which in turn allows these life forms to survive and prosper.

Intelligence is built through the communication of information. Quantum physics suggests that all objects are connected in some way, and that they exist through this exchange of information. Light, in the form of photons, that travels from the sun to our planet carries energy but also drives this sharing of information. These photons allow us to see and collect information. The energy from the sun ensures that our ecosystem on Earth can be sustained and the light that it provides allows us to learn about our environment. Life cannot exist without a combination of energy and information. Biology uses these ingredients to build intelligence. I have come to understand that energy, information, communication and intelligence are all intimately connected.

Our much earlier ancestors produced stone tools and controlled fire. With larger brains, Homo sapiens quickly developed better tools, domesticated animals and started to build a modern society. The anthropogenic capture of new energy sources was a major breakthrough, and the coming of the steam age drove an explosion in technological advances and a take-off in global economic activity.

Humans are very good at developing tools, and artificial intelligence is just another new tool. It will allow us to understand more from the information that is all around us. AI will help us learn much more about ourselves and our environment. It will let us dig deeper into new sources of data to find more information that we can then convert to knowledge. From this new knowledge we can build more intelligence.

Artificial intelligence is a product of our human evolution; it is a technology that has been developed by humans using a combination of our intelligence and our unique tool-building capabilities. AI can't think in the way that we might describe for ourselves, and its purpose is set by us; it doesn't have free will.

In *The Child's Conception of the World*, written in 1929,[11] the Swiss psychologist Jean Piaget said, 'The child begins by seeing purpose everywhere.' Only later do we start to differentiate between the purpose of the things themselves (what we call animism) and the purpose of the makers of things (artificialism). However, our innate childlike view remains part of our human subconscious and causes us to view with suspicion any new devices that appear to display some complex, self-driven purpose. To overcome this innate misunderstanding, we must learn more about this new thing called artificial intelligence. We can then start to understand that its behaviour is artificial, not animistic. The apparent purpose that we observe is embedded by the human makers of this complex tool. Responsible AI must therefore place controls over these human makers.

I believe that information, like oxygen, water, energy and communication, is a key driving force of nature. The communication of information delivers understanding about our environment, which in turn allows us to organize and build structure. Our quest, as humans, for more information and for more understanding is driven by our overarching biological purpose, which is to increase our intelligence.

To return to our definition: 'Intelligence is the ability to gather and use information, in order to adapt and survive.' AI can help us achieve this purpose, but it will continue to become stronger. This brings responsibility, and we must ensure that it is used as a force for good. AI can help us, but importantly it will also allow us to help all the other life forms on our planet. We form part of a single biosphere, and biology shows us that we all originate from a common microbial source. We think of ourselves as distinct and different, superior even, but we share a deep interconnectedness with all other biological life.

Life on our planet constantly surprises me. The more I learn, the more I realize that we are surrounded by the most amazing, complex and intelligent biological machines – machines that we still barely understand. Humans can self-actualize, and so we are able to approach problems with understanding, recognizing our faults and limitations, and hopefully able to accept who we are. We must try to use our superior intelligence with humility and curiosity, to better understand our environment so that we can maintain our world. It is not here just for us.

We are here to pass on information and knowledge to our children and to create a safe environment for all. But we also have a much deeper responsibility: we hold all the cards – the world is in our hands. Evolution has brought us to this elevated position of wholesale control. Our distant evolutionary cousins, which include the house martins that visit my house each year, the bees with their waggle dance and even the colour-blind butterflies – in fact every living organism on this planet, who we are all distantly related to – are *all* depending on us.

AND FINALLY . . .

Throughout the pages of this book, I have tried to show how a set of technological innovations have now allowed us to develop artificial intelligence. AI is a new tool and we have seen how it can have a positive impact on our world, helping us solve problems that were previously out of reach. However, we have also seen that AI is perhaps the most powerful tool that we have ever created, and so we must regulate the way in which AI systems are both developed and applied. The machine is directed, and its purpose is set by developers and organizations. We need new institutions that have enough knowledge to help us regulate them well.

As you have seen, our human intelligence is very different from the artificial intelligence that we can build in a machine. Although humans are biological machines, we are so complex that we do not yet understand all the details of how we work – or even understand how the most basic biological creatures and plants really work.

Our AI systems are still far behind our own very special human intelligence. AI does not think in the same way as humans; instead, artificial intelligence offers something different: a way for us to augment our human intelligence and a way for us to help solve important problems. We must embrace the coming AI revolution.

REFERENCES

INTRODUCTION

1 'Turing's Lecture to the London Mathematical Society on 20 February 1947' in S. Barry Cooper and Jan van Leeuwen, *Alan Turing: His Work and Impact* (Elsevier Science, 2013), pp. 481–97

2 Medeiros, João, 'The science behind Chris Froome and Team Sky's Tour de France preparations', *Wired*, 30 June 2016, https://www.wired.co.uk/article/tour-de-france-science-behind-team-sky

3 Mori, Masahiro, 'The Uncanny Valley: The Original Essay', translated by Karl F. MacDorman and Norri Kageki, *IEEE Robotics & Automation Magazine*, 19: 2 (2012)

4 Clayton, James, 'Meta's chatbot says the company "exploits people"', BBC News, 11 August 2022, https://www.bbc.co.uk/news/technology-62497674

5 'Turing's Lecture to the London Mathematical Society on 20 February 1947', op. cit.

PART 1: HOW AI BECAME POSSIBLE

1 Adams, Douglas, *The Salmon of Doubt: Hitchhiking the Galaxy One Last Time* (William Heinemann, 2002)

CHAPTER 1. THE AI REVOLUTION HAS ALREADY STARTED

1 Turing, A. M., 'Computing machinery and intelligence', *Mind*, 59: 236 (1950), pp. 433–60

2 Bender, Emily M.; Gebru, Timnit; McMillan-Major, Angelina et al., 'On the dangers of stochastic parrots: Can language models be too big?', *Proceedings of the 2021 ACM Conference on Fairness, Accountability, and Transparency*, 2021, pp. 610–623

3 Tiku, Nitasha, 'The Google engineer who thinks the company's AI has come to life', *Washington Post*, 11 June 2022

4 Katz, Daniel M.; Bommarito, Michael J.; Shang, Gao et al., 'GPT-4 passes the bar exam', SSRN, 15 March 2023

CHAPTER 2. INTELLIGENT MACHINES

1 Bianconi, Eva; Piovesan, Allison; Facchin, Federica et al., 'An estimation of the number of cells in the human body', *Annals of Human Biology*, 40: 6 (2013), pp. 463–71

2 Comfort, Nathaniel, 'We are the 98%', *Nature*, 520 (2015), pp. 615–16

3 Parkin, Simon, 'The Space Invader', *New Yorker*, 17 October 2013, https://www.newyorker.com/tech/annals-of-technology/the-space-invader

4 Overney, Leila S.; Blanke, O.; and Herzog, M. H., 'Enhanced temporal but not attentional processing in expert tennis players', *PLOS ONE*, 3: 6 (2008), e2380

5 Moravec, Hans, *Mind Children: The Future of Robot and Human Intelligence* (Harvard University Press, 1988)

6 Gladwell, Malcolm, *Outliers: The Story of Success* (Little, Brown, New York, 2008)

7 Mnih, Volodymyr; Kavukcuoglu, Koray; Silver, David et al., 'Human-level control through deep reinforcement learning', *Nature*, 518 (2015), pp. 529–33

8 Sender, Ron, and Milo, Ron, 'The distribution of cellular turnover in the human body', *Nature Medicine*, 27 (2021), pp. 45–8

9 Paturel, Amy, 'The benefits of sleep for brain health', *Brain & Life*, February/March 2014

CHAPTER 3. THE BIRTH OF AI

1 'Cats and the Internet', Wikipedia, https://en.wikipedia.org/wiki/Cats_and_the_Internet

2 Le, Quoc V.; Ranzato, Marc'Aurelio; Monga, Rajat et al., 'Building high-level features using large scale unsupervised learning', arXiv, 29 December 2011, https://arxiv.org/abs/1112.6209

3 McCarthy, J.; Minsky, M. L.; Rochester, N. et al., 'A Proposal for the Dartmouth Summer Research Project on Artificial Intelligence', 1955, http://raysolomonoff.com/dartmouth/boxa/dart564props.pdf

4 Shannon, Claude E., 'Programming a computer for playing chess', in David Levy, ed., *Computer Chess Compendium* (Springer, New York, 1988)

5 Gödel, Kurt, 'Über formal unentscheidbare Sätze der Principia Mathematica und verwandter Systeme I' (1931), in Solomon Feferman, ed., *Kurt Gödel: Collected Works, Vol. I – Publications 1929–1936* (Oxford University Press, New York, 1986), pp. 144–95. The original German with a facing English translation, preceded by an introductory note by Stephen Cole Kleene.

6 Turing, A. M. 'On computable numbers, with an application to the Entscheidungsproblem', *Proceedings of the London Mathematical Society*, 42: 1 (1937), pp. 230–65

7 For Steve Jobs talking about computers as bicycles for the mind, see https://www.youtube.com/watch?v=4x8wTj-n33A; for the *Scientific American* article Jobs refers to see Wilson, S. S., 'Bicycle technology', *Scientific American*, 228: 3 (1973), pp. 81–91

CHAPTER 4. THE TECHNOLOGY THAT BUILT AI – PART 1: ELECTRONIC COMPUTERS AND LEARNING TO SEE

1 Copeland, B. Jack, ed., *Colossus: The Secrets of Bletchley Park's Codebreaking Computers* (Oxford University Press, Oxford, 2010), Chapter 6

2 Ibid., p. 75

3 Ibid., Chapter 6

4 Ibid., Chapter 11

5 Krisciunas, Kevin, and Carona, Don W., 'At what distance can the human eye detect a candle flame?', arXiv:1507.06270 [astro-ph.IM]

6 Potter, Mary C.; Wyble, Brad; Hagmann, Carl Erick et al., 'Detecting meaning in RSVP at 13 ms per picture', *Attention, Perception, & Psychophysics*, 76: 2 (2014), pp. 270–9

7 Grady, Cheryl L.; McIntosh, Anthony R.; Rajah, M. Natasha et al., 'Neural correlates of the episodic encoding of pictures and words', *Proceedings of the National Academy of Sciences*, 95: 5 (1998), pp. 2703–8

8 Sheth, Bhavin R.; Sharma, Jitendra; Chenchal Rao, S. et al., 'Orientation maps of subjective contours in visual cortex', *Science*, 274: 5295 (1996), pp. 2110–15

9 Hubel, David H., and Wiesel, Torsten N., 'Effects of monocular deprivation in kittens', *Naunyn-Schmiedebergs Archiv für experimentelle Pathologie und Pharmakologie*, 248 (1964), pp. 492–7

CHAPTER 5. THE TECHNOLOGY THAT BUILT AI – PART 2: SEMICONDUCTORS, SOFTWARE AND ATTENTION

1 Moore, G. E., 'Cramming more components onto integrated circuits, reprinted from *Electronics*, volume 38, number 8, April 19, 1965, pp. 114 ff.' in *IEEE Solid-State Circuits Society Newsletter*, 11: 3 (2006), pp. 33–5

2 Toole, Betty Alexandra, *Ada, the Enchantress of Numbers: Prophet of the Computer Age* (Strawberry Press, 1998), p. 38

3 Hollings, Christopher; Martin, Ursula; and Rice, Adrian, *Ada Lovelace: The Making of a Computer Scientist* (Bodleian Library, University of Oxford, 2018)

4 Simonite, Tom, 'Short Sharp Science: Celebrating Ada Lovelace: the world's first programmer', *New Scientist*, 24 March 2009

5 Dingfelder, S. F., 'Can rats reminisce?', *Monitor on Psychology*, 38: 7 (2007), p. 26

6 Tulving, Endel, 'How many memory systems are there?', *American Psychologist*, 40: 4 (1985), pp. 385–98

7 Vaswani, Ashish; Shazeer, Noam; Parmar, Niki et al., 'Attention is all you need', *Advances in Neural Information Processing Systems* (2017), pp. 5998–6008

8 Jumper, John; Evans, Richard; Pritzel, Alexander et al., 'Highly accurate protein structure prediction with AlphaFold', *Nature*, 596: 7873 (2021), pp. 583–9

9 Burley, Stephen K.; Berman, Helen M.; Duarte, José M. et al., 'Protein

Data Bank: A comprehensive review of 3D structure holdings and worldwide utilization by researchers, educators, and students', *Biomolecules*, 12: 10 (2022), p. 1425

10 'AI and compute', OpenAI, https://openai.com/blog/ai-and-compute/

11 Gilbert, Lynn, and Moore, Gaylen, 'Women of Wisdom – Talks with Women Who Shaped Our Times: Grace Murray Hopper', 1981, https://www.vassar.edu/stories/2017/assets/images/170706-legacy-of-grace-hopper-hopperpdf.pdf

12 Strawn, George, and Strawn, Candace, 'Grace Hopper: Compilers and Cobol', *IT Professional*, 17: 1 (2015), pp. 62–4

13 Grace Hopper MIT Lincoln Laboratory lecture on the future of computing, https://www.youtube.com/watch?v=ZR0ujwlvbkQ

14 Andreessen, Marc, 'Why software is eating the world', Andreessen Horowitz blog, https://future.a16z.com/software-is-eating-the-world/

CHAPTER 6. THE TECHNOLOGY THAT BUILT AI – PART 3: GETTING CREATIVE, GETTING CONNECTED, AND GETTING MORE INFORMATION

1 Duan, Yan; Andrychowicz, Marcin; Stadie, Bradly C. et al., 'One-shot imitation learning', arXiv, 21 March 2017, https://arxiv.org/abs/1703.07326

2 Ho, Jonathan; Jain, Ajay; and Abbeel, Pieter, 'Denoising diffusion probabilistic models', arXiv, 19 June 2020, https://arxiv.org/abs/2006.11239

3 Shields, Rob, 'Cultural topology: The seven bridges of Königsburg, 1736', *Theory, Culture & Society*, 29: 4–5 (2012), pp. 43–57

4 Euler, Leonhard, 'Solutio problematis ad geometriam situs pertinentis' (1736), https://archive.org/details/commentariiacade08impe/page/128/mode/2up

5 'Great Elephant Census', Paulallen.com, https://paulallen.com/Exploration/Great-Elephant-Census.aspx

6 Hutter, Marcus, 'A theory of universal artificial intelligence based

on algorithmic complexity', arXiv, 2 April 2000, https://arxiv.org/abs/cs/0004001; Hutter, Marcus, 'One decade of Universal Artificial Intelligence', arXiv, 28 February 2012, https://arxiv.org/abs/1202.6153; Legg, Shane, 'Machine Super Intelligence', PhD thesis, Faculty of Informatics, University of Lugano, 2008

7 Hutter, 'A theory of universal artificial intelligence', p. 9

8 Silver, David; Huang, Aja; Maddison, Chris J. et al., 'Mastering the game of Go with deep neural networks and tree search', *Nature*, 529 (2016), pp. 484–9

9 Silver, David; Schrittwieser, Julian; Simonyan, Karen et al., 'Mastering the game of Go without human knowledge', *Nature*, 550 (2017), pp. 354–9

10 Silver, David; Hubert, Thomas; Schrittwieser, J. et al., 'A general reinforcement learning algorithm that masters chess, shogi, and Go through self-play', *Science*, 362: 6419 (2018), pp. 1140–4

11 Hume, David, *Philosophical Essays Concerning Human Understanding*, first edition (A. Millar, London, 1748)

12 'An essay towards solving a problem in the doctrine of chances. By the late Rev. Mr. Bayes, F. R. S. communicated by Mr. Price, in a letter to John Canton, A. M. F. R. S', Royal Society, https://royalsocietypublishing.org/doi/10.1098/rstl.1763.0053

13 Solomonoff, R. J., 'A formal theory of inductive inference. Part I', *Information and Control*, 7: 1 (1964), pp. 1–22; 'A formal theory of inductive inference. Part II', *Information and Control*, 7: 2 (1964), pp. 224–54

14 'Sir Tim Berners-Lee: Father of the World Wide Web', American Academy of Achievement interview, 22 June 2007, https://achievement.org/achiever/sir-timothy-berners-lee/#interview

15 'Remarks given by Vice President Gore at the Superhighway Summit, UCLA', 11 January 1994

16 Cisco Annual Internet Report, https://www.cisco.com/c/en/us/solutions/executive-perspectives/annual-internet-report/index.html

CHAPTER 7. CLAUDE SHANNON, THE FATHER OF OUR INFORMATION AGE

1 'Claude Shannon: Reluctant father of the digital age', *MIT Technology Review*, 1 July 2001, https://www.technologyreview.com/2001/07/01/235669/claude-shannon-reluctant-father-of-the-digital-age/

2 Cheshire, Tom, 'How to count to 1,023 on your fingers', *Wired*, 4 June 2010, https://www.wired.co.uk/article/how-to-count-to-1023-on-your-fingers

3 Shannon, Claude E., 'A symbolic analysis of relay and switching circuits', *Transactions of the American Institute of Electrical Engineers*, 57: 12 (1938), pp. 713–23

4 Shannon, C. E., 'A mathematical theory of communication', *Bell System Technical Journal*, 27: 3 (1948), pp. 379–423

5 Sterling, Peter, and Laughlin, Simon, *Principles of Neural Design* (MIT Press, 2017)

6 Wrangham, Richard, *Catching Fire: How Cooking Made Us Human* (Basic Books, 2009)

7 Shannon, Claude Elwood, 'An Algebra for Theoretical Genetics', PhD thesis, Department of Mathematics, Massachusetts Institute of Technology, 1940

8 Soni, Jimmy, and Goodman, Rob, *A Mind at Play: How Claude Shannon Invented the Information Age* (Simon & Schuster, 2017)

PART 2: HOW IS AI DIFFERENT FROM HUMAN INTELLIGENCE?

1 'U of T computer scientist takes international prize for groundbreaking work in AI', *U of T News*, 18 January 2017

CHAPTER 8. WHAT IS INTELLIGENCE?

1 Herculano-Houzel, S. 'The remarkable, yet not extraordinary, human brain as a scaled-up primate brain and its associated cost', *Proceedings of the National Academy of Sciences*, 109 S1 (2012), pp. 10661–8

2 McCulloch, Warren S., and Pitts, Walter, 'A logical calculus of the ideas immanent in nervous activity', *Bulletin of Mathematical Biophysics*, 5 (1943), pp. 115–133

CHAPTER 9: MORE INTELLIGENCE

1 Turing, Alan (voiced by Mark Gatiss), 'Can digital computers think?', lecture broadcast on the BBC Home Service, 15 May 1951, https://www.bbc.co.uk/archive/alan_turing_can_computers_think/zb8rkhv

2 Einstein, Albert, and Infeld, Leopold (1938), *The Evolution of Physics: The Growth of Ideas from Early Concepts to Relativity and Quanta* (Simon & Schuster, New York, 1983)

3 Moore, G. E., 'Cramming more components onto integrated circuits', reprinted from *Electronics*, volume 38, number 8, April 19, 1965, pp. 114 ff. in *IEEE Solid-State Circuits Society Newsletter*, 11: 3 (2006), pp. 33–5

4 Turing, A. M. 'On computable numbers, with an application to the Entscheidungsproblem', *Proceedings of the London Mathematical Society*, 42: 1 (1937), pp. 230–65

5 Einstein and Infeld, op. cit.

6 For a comprehensive list of definitions of intelligence, see Legg, Shane, and Hutter, Marcus, 'A Collection of Definitions of Intelligence', arXiv: https://arxiv.org/pdf/0706.3639.pdf

7 Lovelock, J. E., 'Gaia as seen through the atmosphere', *Atmospheric Environment*, 6: 8 (1972), pp. 579–80

8 Cartier-Bresson, H.; Matisse, H.; and Tériade, E., *The Decisive Moment* (Simon & Schuster in collaboration with Éditions Verve of Paris, 1952)

9 McGurk, Harry, and MacDonald, John, 'Hearing lips and seeing voices', *Nature*, 264 (1976), pp. 746–8

10 Tang, Yu-Chong; Zhou, Cheng-Li; Chen, Xiao-Ming et al., 'Visual and olfactory responses of seven butterfly species during foraging', *Journal of Insect Behavior*, 26 (2013), pp. 387–401

11 Herger, Mario, '2022 Disengagement Report from California', The Last Driver License Holder Has Already Been Born, 17 February 2023, https://thelastdriverlicenseholder.com/2023/02/17/2022-disengagement-report-from-california/

12 Heisenberg, Werner, translated by Eckart, Carl, and Hoyt, Frank C., *The Physical Principles of the Quantum Theory* (Dover, 1955)

13 Einstein, Albert, 'Jedenfalls bin ich überzeugt, daß der nicht würfelt' (1926) in *The Born–Einstein Letters: Correspondence Between Albert Einstein and Max and Hedwig Born 1916–1955*, translated by Irene Born (Walker and Company, New York, 1971)

14 Soni, Jimmy, and Goodman, Rob, *A Mind at Play: How Claude Shannon Invented the Information Age* (Simon & Schuster, 2017))

15 Gleick, James, *The Information: A History, a Theory, a Flood* (Pantheon, 2011)

16 Schrödinger, Erwin, *What is Life? & Other Scientific Essays*, based on lectures delivered under the auspices of the Dublin Institute for Advanced Studies at Trinity College, Dublin, in February 1943 (Doubleday, New York, 1944)

CHAPTER 10. CONSCIOUSNESS

1 Riley, J. R.; Greggers, U.; Smith, A. et al., 'The flight paths of honeybees recruited by the waggle dance', *Nature*, 435 (2005), pp. 205–207

2 Kahneman, Daniel, *Thinking, Fast and Slow* (Farrar, Straus and Giroux, 2011)

3 Schrödinger, Erwin, *What is Life? & Other Scientific Essays*, based on lectures delivered under the auspices of the Dublin Institute for Advanced Studies at Trinity College, Dublin, in February 1943 (Doubleday, New York, 1944)

4 Marr, David, *Vision: A Computational Investigation into the Human Representation and Processing of Visual Information* (W. H. Freeman, San Francisco, 1982)

5 Baars, Bernard J., *A Cognitive Theory of Consciousness* (Cambridge University Press, Cambridge, Massachusetts, 1988); Baars, Bernard J., *In the Theater of Consciousness: The Workspace of the Mind* (Oxford University Press, New York, 1997); Baars, Bernard J., 'The conscious access hypothesis: Origins and recent evidence', *Trends in Cognitive Sciences*, 6: 1 (2002), pp. 47–52

6 Tononi, Giulio, 'Integrated information theory', Scholarpedia, 2015, https://doi.org/10.4249/scholarpedia.4164

7 Bender, Emily M.; Gebru, Timnit; McMillan-Major, Angelina et al., 'On the dangers of stochastic parrots: Can language models be too big?', Proceedings of the 2021 ACM Conference on Fairness, Accountability, and Transparency, 2021, pp. 610–623

8 Koch, Christof, 'What is consciousness?', *Nature*, 557 (2018), S8-S12

9 Chalmers, D. J., 'Facing up to the problem of consciousness', *Journal of Consciousness Studies*, 2 (1995), pp. 200–219

10 Shannon, C. E., 'A mathematical theory of communication', *Bell System Technical Journal*, 27: 3 (1948), pp. 379–423

11 Schrödinger, op. cit.

12 Asimov, Isaac, *I, Robot* (Fawcett Publications, 1950)

13 *The Social Dilemma*, directed by Jeff Orlowski-Yang, Netflix, netflix.com/title/81254224

14 Moravec, H. P., *Mind Children: The Future of Robot and Human Intelligence* (Harvard University Press, 1998)

CHAPTER 11. AN ULTRA-INTELLIGENT MACHINE

1 Good, Irving John, 'Speculations concerning the first ultra-intelligent machine', *Advances in Computers*, 6 (1966), pp. 31–88

2 Kurzweil, Ray, *The Singularity Is Near: When Humans Transcend Biology* (Penguin Books, 2005)

3 Zimmer, Carl, '100 trillion connections: New efforts probe and map the brain's detailed architecture', *Scientific American*, 1 January 2011

4 Roy, Brandon C.; Frank, Michael C.; DeCamp, Philip et al., 'Predicting the birth of a spoken word', *Proceedings of the National Academy of Sciences*, 112: 41 (2015), pp. 12663–82

5 For a list of predictions on the compute performance of the brain, see 'Brain performance in FLOPS', AI Impacts, https://aiimpacts.org/brain-performance-in-flops/

6 'Huge "foundation models" are turbo-charging AI progress', *Economist*,

11 June 2022, https://www.economist.com/interactive/briefing/2022/06/11/huge-foundation-models-are-turbo-charging-ai-progress

CHAPTER 12. IS A SINGULARITY EVENT POSSIBLE?

1 Kempes, Christopher P.; Wolpert, David; Cohen, Zachary et al., 'The thermodynamic efficiency of computations made in cells across the range of life', *Philosophical Transactions of the Royal Society A*, 375: 2109 (2017)

2 Landauer, Rolf, 'Irreversibility and heat generation in the computing process', *IBM Journal of Research and Development*, 5: 3 (1961), pp. 183–91

3 Dennard, R. H.; Gaensslen, F. H.; Yu, Hwa-Nien et al., 'Design of ion-implanted MOSFET's with very small physical dimensions', *IEEE Journal of Solid-State Circuits*, 9: 5 (1974), pp. 256–68

4 Bassett, Danielle S.; Greenfield, Daniel L.; Meyer-Lindenberg, Andreas et al., 'Efficient physical embedding of topologically complex information processing networks in brains and computer circuits', *PLOS Computational Biology*, 6: 4 (2010)

5 Amara's law: 'Roy Amara 1925–2007, American futurologist', *Oxford Essential Quotations*, Vol. 1, fourth edition (Oxford University Press, 2016), https://doi.org/10.1093/acref/9780191826719.001.0001

6 Lanzerotti, M. Y.; Fiorenza, G.; and Rand, R. A., 'Microminiature packaging and integrated circuitry: The work of E. F. Rent, with an application to on-chip interconnection requirements' in *IBM Journal of Research and Development*, 49: 4.5 (2005), pp. 777–803

7 Doherty, Sally, 'Designing the Colossus Mk2 IPU: Simon Knowles at Hot Chips 2021', Graphcore, 25 August 2021, https://www.graphcore.ai/posts/designing-the-colossus-mk2-ipu-simon-knowles-at-hot-chips-2021; https://www.graphcore.ai

8 For a deeper academic understanding of quantum computers see: Bennett, C. H., 'Logical reversibility of computation', *IBM Journal of Research and Development*, 17: 6 (1973), pp. 525–32; Benioff, Paul, 'Quantum mechanical models of Turing machines that dissipate no energy', *Physical Review Letters*, 48: 1581 (1982), pp. 1581–1585; Shor, Peter W., 'Algorithms for quantum computation: discrete logarithms and factoring', *Proceedings 35th Annual Symposium on Foundations of Computer Science* (1994), pp. 124–34

9 O'Brien, Jeremy, 'Building the world's first useful quantum computer', PsiQuantum, 6 April 2020, https://psiquantum.com/news/building-the-worlds-first-useful-quantum-computer

10 Good, Irving John, 'Speculations concerning the first ultraintelligent machine', *Advances in Computers*, 6 (1966), pp. 31–88

11 Ionkov, Latchesar, and Settlemyer, Bradley, 'DNA: The Ultimate Data-Storage Solution', *Scientific American*, 28 May 2021, https://www.scientificamerican.com/article/dna-the-ultimate-data-storage-solution/

PART 3: DO WE UNDERSTAND AI'S INCREDIBLE POTENTIAL?

1 Minsky, Marvin, *The Society of Mind* (Simon & Schuster, 1986)

CHAPTER 13. AI AND THE ENVIRONMENT

1 Integrated Carbon Observation System (ICOS), https://www.icos-cp.eu; Ritchie, Hannah, and Roser, Max, 'CO2 emissions', Our World in Data, https://ourworldindata.org/co2-emissions

2 Gates, Bill, *How to Avoid a Climate Disaster* (Allen Lane, 2021)

3 Troutman, Keri, 'Advancing new battery design with deep learning', Berkeley Research, 6 April 2022, https://vcresearch.berkeley.edu/news/advancing-new-battery-design-deep-learning

4 Eddington, A. S., 'The internal constitution of the stars', *Nature*, 106 (1920), pp. 14–20

5 'The way ahead for fusion', *Nature Physics*, 16: 9 (2020), p. 889

6 Degrave, Jonas; Felici, Federico; Buchli, Jonas et al., 'Magnetic control of tokamak plasmas through deep reinforcement learning', *Nature*, 602 (2022), pp. 414–19

7 Kusuma, Julius, and Sudhalkar, Amruta, 'Green concrete: Using AI to reduce concrete's carbon footprint', Tech at Meta, https://tech.facebook.com/engineering/2022/04/sustainable-concrete/

8 Smil, Vaclav, *How the World Really Works: A Scientist's Guide to Our Past, Present and Future* (Penguin, 2022)

CHAPTER 14. AI IN EDUCATION

1 'More than half of children and youth worldwide "not learning" – UNESCO', United Nations, 21 September 2017, https://www.un.org/sustainabledevelopment/blog/2017/09/more-than-half-of-children-and-youth-worldwide-not-learning-unesco/

2 Hao, Karen, 'China has started a grand experiment in AI education. It could reshape how the world learns', *MIT Technology Review*, 1 August 2019

3 'Goal 4: Ensure inclusive and equitable quality education and promote lifelong learning opportunities for all', United Nations, https://unstats.un.org/sdgs/report/2019/goal-04/

4 Rid, Thomas, 'Five Days in Class with ChatGPT', Alperovitch Institute blog, 22 January 2023

5 Gates, Bill, 'The Age of AI has begun', GatesNotes blog, 21 March 2023

CHAPTER 15. THE AI HEALTHCARE REVOLUTION

1 Topol, Eric, *Deep Medicine: How Artificial Intelligence Can Make Healthcare Human Again* (Basic Books, 2019)

2 'Life expectancy up: The best places to be born, work, and retire in India', *Times of India*, 13 June 2022

3 Hern, Alex, 'Royal Free breached UK data law in 1.6m patient deal with Google's DeepMind', *Guardian*, 3 July 2017, https://www.theguardian.com/technology/2017/jul/03/google-deepmind-16m-patient-royal-free-deal-data-protection-act

4 'Average cost of developing a new drug could be up to $1.5 billion less than pharmaceutical industry claims', London School of Economics and Political Science, 3 March 2020, https://www.lse.ac.uk/News/Latest-news-from-LSE/2020/c-March-20/Average-cost-of-developing-a-new-drug-could-be-up-to-1.5-billion-less-than-pharmaceutical-industry-claims

5 Mullard, Asher, 'The drug-maker's guide to the galaxy', *Nature*, 549 (2017), pp. 445–7

6 Nawrat, Allie, 'Exscientia Q&A: leveraging AI to create bispecific small molecule drugs', *Pharmaceutical Technology*, 31 October

2009, https://www.pharmaceutical-technology.com/features/exscientia-ai-bispecific-small-molecule-drugs/

7 Zhavoronkov, Alex; Ivanenkov, Yan A.; Aliper, Alex et al., 'Deep learning enables rapid identification of potent DDR1 kinase inhibitors', *Nature Biotechnology*, 37 (2019), pp. 1038–40

8 McCann, Chandler, 'Worldwide water access: Tapping into a well of data', DataRobot, 16 July 2019, https://www.datarobot.com/blog/worldwide-water-access-tapping-into-a-well-of-data/

9 Goedde, Lutz; Ooko-Ombaka, Amandla; and Pais, Gillian, 'Winning in Africa's agricultural market', McKinsey & Company, https://www.mckinsey.com/industries/agriculture/our-insights/winning-in-africas-agricultural-market

10 'Water scarcity: Addressing the growing lack of available water to meet children's needs', Unicef, https://www.unicef.org/wash/water-scarcity

CHAPTER 16. THE CHALLENGES OF AI

1 From 'A New Frontier: National Security, Artificial Intelligence, and Misinformation', a talk by Eric Schmidt at the Aspen Security Forum, 22 July 2022, https://www.youtube.com/watch?v=ryuaQwcKZwY

2 'Pause giant AI experiments: An open letter', Future of Life Institute, 22 March 2023, https://futureoflife.org/open-letter/pause-giant-ai-experiments/

3 Holland, Tom, *Dominion: The Making of the Western Mind* (Basic Books, 2021)

4 Arthur, Charles, 'Tech giants may be huge, but nothing matches big data', *Guardian*, 23 April 2014, https://www.theguardian.com/technology/2013/aug/23/tech-giants-data

5 Metz, Cade, 'Google patents search that tracks your moves', *Register*, 27 July 2020, https://www.theregister.com/2010/07/27/google_patents_mouse_movement_search_tweaks/

6 Dunning, David, 'Chapter five – The Dunning–Kruger effect: On being ignorant of one's own ignorance', *Advances in Experimental Social Psychology*, 44 (2011), pp. 247–96

7 Kosinski, Michal; Stillwell, David; and Graepel, Thore, 'Private traits and attributes are predictable from digital records of human behavior', *Proceedings of the National Academy of Sciences*, 110: 15 (2013), pp. 5802–5

8 Schuller, Dagmar, and Schuller, Björn W., 'The age of artificial emotional intelligence', *Computer*, 51: 9 (2018), pp. 38–46

9 Ackerman, Sandra, *Discovering the Brain* (National Academies Press, 1992)

10 Kaplan, Matt, 'How to laugh away stress', *Nature* (2008), https://doi.org/10.1038/news.2008.741

11 Bellace, Matt, *A Better High: Laugh, Help, Run, Love . . . and Other Ways to Get Naturally High!* (Wyatt-MacKenzie, 2012)

12 Miller, Daniel; Abed Rabho, Laila; Awondo, Patrick et al., *The Global Smartphone: Beyond a Youth Technology* (UCL Press, 2021), pp. i–iv

13 Picard, R. W., *Affective Computing* (MIT Press, 1997)

14 *Her*, film directed by Jonze, Spike (Warner Bros. Pictures, 2013)

15 Lonas, Lexi, 'Facebook formula gave anger five times weight of likes, documents show', *Hill*, 26 October 2021, https://thehill.com/policy/technology/578548-facebook-formula-gave-anger-five-times-weight-of-likes-documents-show/

16 'Universal Declaration of Human Rights' (Paris, 1948), 217 (III] A, http://www.un.org/en/universal-declaration-human-rights

17 Autor, David H., 'Why are there still so many jobs? The history and future of workplace automation', *Journal of Economic Perspectives*, 29: 3 (2015), pp. 3–30

18 '63 per cent of companies consider Excel a vital accounting tool', *Financial Post*, 29 April 2021, https://financialpost.com/personal-finance/business-essentials/63-per-cent-of-companies-consider-excel-a-vital-accounting-tool

19 'Driving impact at scale from automation and AI', McKinsey, February 2019, https://www.mckinsey.com/~/media/McKinsey/Business%20Functions/McKinsey%20Digital/Our%20Insights/Driving%20impact%20at%20scale%20from%20automation%20and%20AI/Driving-impact-at-scale-from-automation-and-AI.ashx

20 Smith, Adam, *An Inquiry into the Nature and Causes of the Wealth of Nations* (W. Strahan and T. Cadell, 1776)

21 *Under Pressure: The Squeezed Middle Class* (OECD Publishing, 2019)

22 'Letter dated 8 March 2021 from the panel of experts on Libya established pursuant to a resolution 1973 (2011) addressed to the president of the Security Council', United Nations Security Council, 2021, https://documents-dds-ny.un.org/doc/UNDOC/GEN/N21/037/72/PDF/N2103772.pdf?OpenElement

23 Farge, Emma, 'U.N. talks adjourn without deal to regulate "killer robots"', Reuters, 17 December 2021, https://www.reuters.com/world/un-talks-adjourn-without-deal-regulate-killer-robots-2021-12-17/

CHAPTER 17: RESPONSIBLE AI

1 Frantz, Laurent A. F.; Bradley, Daniel G.; Larson, Greger et al., 'Animal domestication in the era of ancient genomics', *Nature Reviews: Genetics*, 21 (2020), pp. 449–60

2 'In their own words', from *Century Magazine*, September 1908, reproduced at Wright Brothers Aeroplane Company, https://www.wright-brothers.org/History_Wing/Wright_Story/Showing_the_World/Tragedy_at_Fort_Myer/Wright_Brothers_Aeroplane.htm

3 Michael Way, '"What I cannot create, I do not understand"', *Journal of Cell Science*, 130: 18 (2017), pp. 2941–2

4 Jarvis Thomson, Judith, 'The trolley problem', *Yale Law Journal*, 94: 6 (1985), pp. 1395–1415

5 Millar, Jason, 'You should have a say in your robot car's code of ethics', *Wired*, 2 September 2014, https://www.wired.com/2014/09/set-the-ethics-robot-car/

6 'An essay towards solving a problem in the doctrine of chances. By the late Rev. Mr. Bayes, F. R. S. communicated by Mr. Price, in a letter to John Canton, A. M. F. R. S', Royal Society, https://royalsocietypublishing.org/doi/10.1098/rstl.1763.0053

CHAPTER 18: HOW TECHNOLOGY REVOLUTIONS HAPPEN

1 'Wanamaker's', Wikipedia, https://en.wikipedia.org/wiki/Wanamaker%27s

2 'Thomas Newcomen Engine', Devon Museums, https://www.devonmuseums.net/Thomas-Newcomen-Engine/Devon-Museums

3 'Watt steam engine', Wikipedia, https://en.wikipedia.org/wiki/Watt_steam_engine

4 'Richard Trevithick', Wikipedia, https://en.wikipedia.org/wiki/Richard_Trevithick

5 'Liverpool and Manchester Railway', Wikipedia, https://en.wikipedia.org/wiki/Liverpool_and_Manchester_Railway

6 'Michael Faraday (1791–1876)', Royal Institution, https://www.rigb.org/explore-science/explore/person/michael-faraday-1791-1867

7 'Holborn Viaduct power station', Wikipedia, https://en.wikipedia.org/wiki/Holborn_Viaduct_power_station

8 'The Mobile Economy 2022', GSM Association, 2022, https://www.gsma.com/mobileeconomy/wp-content/uploads/2022/02/280222-The-Mobile-Economy-2022.pdf

9 Eveleth, Rose, 'What was the first book ever ordered on Amazon.com?', *Smithsonian Magazine*,17 April 2013, https://www.smithsonianmag.com/smart-news/what-was-the-first-book-ever-ordered-on-amazoncom-24406844/

10 Krishnan, Mekala; Mischke, Jan; and Remes, Jaana, 'Is the Solow Paradox back?', McKinsey Digital, 4 June 2018, https://www.mckinsey.com/capabilities/mckinsey-digital/our-insights/is-the-solow-paradox-back

11 'Developer Nation: Pulse Report', Developer Nation, https://www.developernation.net/developer-reports/de20

CHAPTER 19. MAKING AI WORK FOR US

1 Hawking, Stephen, *Brief Answers to the Big Questions* (John Murray, 2020)

2 Elon Musk, speaking at the MIT Aeronautics and Astronautics department's Centennial Symposium in October 2014: 'I think we should be very careful about artificial intelligence. If I were to guess, like,

what our biggest existential threat is, it's probably that. So we need to be very careful with the artificial intelligence. Increasingly, scientists think there should be some regulatory oversight, maybe at the national and international level, just to make sure that we don't do something very foolish. With artificial intelligence we are summoning the demon. In all those stories where there's the guy with the pentagram and the holy water, it's like, "Yeah, he's sure he can control the demon." Didn't work out.'; 'AI potentially "more dangerous than nukes," Musk warns', CNBC, 4 August 2014

3 Bostrom, Nick, *Superintelligence: Paths, Dangers, Strategies* (Oxford University Press, Oxford, 2014)

4 Russell, Stuart, '2015: What do you think about machines that think?', Edge, 2015, https://www.edge.org/response-detail/26157

5 Vallance, Chris, 'Meta scientist Yann LeCun says AI won't destroy jobs forever', BBC News, 15 June 2023

6 'Putin: Leader in artificial intelligence will rule the world', CNBC, 4 September 2017, https://www.cnbc.com/2017/09/04/putin-leader-in-artificial-intelligence-will-rule-world.html

7 Lee, Kai-Fu, *AI Superpowers: China, Silicon Valley, and the New World Order* (Houghton Mifflin Harcourt, Boston, Massachusetts, 2018)

8 Webster, Graham; Creemers, Rogier; Triolo, Paul et al., 'Full translation: China's "New Generation Artificial Intelligence Development Plan" (2017)', *New America*, 1 August 2017, https://www.newamerica.org/cybersecurity-initiative/digichina/blog/full-translation-chinas-new-generation-artificial-intelligence-development-plan-2017/

9 Atkinson. S, and Skinner. G, 'What worries the world?', *IPSOS Public Affairs*, 2019

10 'OHCHR Assessment of Human Rights Concerns in the Xinjiang Uyghur Autonomous Region, People's Republic of China', United Nations, 31 August 2022

11 Piaget, Jean, *The Child's Conception of the World* (Kegan Paul, Trench, Trubner & Co., 1929)

GLOSSARY

ARTIFICIAL INTELLIGENCE (AI)

Artificial intelligence describes any type of machine that is built using an electronic computer and can actively learn from information, rather than being told what to do, step by step, in a software program. AI systems will exhibit 'intelligent' characteristics such as reliably recognizing objects in images or understanding language in text.

ARTIFICIAL NEURAL NETWORK

An artificial neural network is the building block now most commonly used for creating artificial-intelligence systems. An artificial neural network attempts to mimic (at a very basic level) the behaviour of biological neural networks found in the brains of animals and insects. Complex artificial neural networks are often termed 'deep neural networks' and many different forms have been developed, including 'convolutional neural networks' and 'transformer'-based artificial neural networks.

ATTENTION

Attention-based AI refers to a machine learning method that learns which parts of language (or other information) to focus attention on during the machine learning process. This type of AI is built using multi-headed attention-based transformer neural networks.

BAYES' THEOREM

Named after the church minister and mathematician Thomas Bayes, Bayes' theorem describes the probability of an event occurring by taking into account prior information or knowledge that may be relevant. For an example, it has been shown that people who regularly smoke are more likely to suffer from lung cancer in later life. Bayes' theorem takes this known information into account when assessing the risk of lung cancer for smokers vs non-smokers, leading to more accurate predictions.

BIAS

Biased data or information disproportionately favours one set of ideas or things at the expense of other ideas or things. An example might be information that isn't representative of real-world demography (for instance, it is estimated that as many as 90 per cent of Wikipedia editors are male, which may bias the information represented there). Bias in the information that is used to train AI systems can cause the AI system to produce biased answers.

BIT

A 'bit' (a contraction of 'binary digit') is the most basic unit of information in computing and digital communications. A bit can only ever be a '1' or a '0', like a switch that is either on or off. Groups of bits can be used to describe numbers and letters, and more complex sequences of bits can be used to describe any information.

BYTE

A byte is a sequence of eight bits. A byte can describe 255 unique numbers or pieces of information.

CENTRAL PROCESSING UNIT (CPU)

A central processing unit is the type of general-purpose computing engine found inside a laptop or mobile phone and is commonly called a 'microprocessor'. CPUs are flexible engines and can perform all types of computing functions, but they may be slower at specific operations – such as complex arithmetic or drawing three-dimensional graphic images on a screen – than a specialized processor such as a graphics processing unit.

CONVOLUTION

A convolution is a type of mathematical operation where one mathematical function describes how a second function is modified to create a third (output) function. To put this more simply, a convolution can be thought of as a form of filtering, like the kind used to recognize the lines and edges around shapes in a picture.

CONVOLUTIONAL NEURAL NETWORK (CNN)

A convolutional neural network is a form of artificial neural network that combines many layers of filtering (or convolution layers) to create an AI system that can recognize objects in images or recognize other information in large amounts of data.

DATA, INFORMATION AND KNOWLEDGE

Data is technically the plural of datum, though it is usually used in the singular to refer to a set of quantities, characters or symbols that describe results, objects or statistics. Data is distinct from information; when data is combined with context to provide useful facts and analysis, it is information. From information, we can build knowledge by understanding the relationships between related pieces of information.

DENNARD SCALING

Named after electronics engineer and IBM researcher Robert Dennard, Dennard scaling is the observation that in semiconductor integrated circuits the power consumption of a transistor will scale in proportion to the area that the transistor occupies. So, as transistors get smaller the power consumption will reduce. The semiconductor industry has now reached the limit of Dennard scaling.

EXAFLOP

A flop, or floating-point operation, is a measure of computing performance. It describes a single complex mathematical operation, such as multiplying two large numbers together, being performed by the machine in one second. An *Exa*flop machine is the term used to describe a machine that is able to perform 1 billion billion flops in a single second. It has been estimated that the human brain may have the same performance as an Exaflop machine.

GENERATIVE AI

Generative AI is a form of artificial intelligence that can generate new information – in the form of text, pictures, music or other media – in response to a language prompt. The term has been popularized by new AI applications such as ChatGPT and Bard, which show an impressive ability to generate new text information, such as a story, from a simple text prompt in ordinary human language.

GENERATIVE PRE-TRAINED TRANSFORMER (GPT)

A generative pre-trained transformer is a form of artificial neural network. GPT is an example of an 'attention'-based technique that learns which parts of language to focus attention on. Through this method the machine can learn the structure of language so that it can analyse a piece of text then generate new text from a simple human-language prompt.

GRAPH NEURAL NETWORK (GNN)

A graph neural network is a form of complex artificial neural network that can be used to describe and understand complex sets of information and relationships. For example, a GNN may understand the connections in a social media network, or the potential connections between buyers and the products in an online store.

GRAPHICS PROCESSING UNIT (GPU)

A graphics processing unit is a specialized form of computing engine that was originally developed to accelerate the rendering of the three-dimensional graphics that you find in computer games. GPUs can perform a large number of complex mathematic operations every second and are now being used to speed up artificial-intelligence machine learning techniques. Even more specialized computing engines are also now being developed to make these techniques ever faster.

INTEGRATED CIRCUIT (IC)

An integrated circuit is an electronic device that combines transistors on a single microchip. A CPU microprocessor is a form of integrated circuit.

Today's ICs are able to combine many billions of transistors, and can perform the function of very complex electronic systems on a single chip that measures just a few centimetres square.

LARGE LANGUAGE MODEL (LLM)

A large language model is a very large artificial neural network that creates AI systems able to understand and generate human language text. As an example, OpenAI has a large language model called GPT-4, which users interact with through the AI language system ChatGPT.

LETHAL AUTONOMOUS WEAPON (LAW)

A lethal autonomous weapon is a type of military weapon system that can independently search for and engage targets based on a set of instructions provided by humans. We need to consider the ethics of designing lethal autonomous weapons that, after launch, could operate without any further human interaction, allowing the weapon to 'decide' whether or not to take a life.

LONG SHORT-TERM MEMORY (LSTM)

A long short-term memory system is a form of artificial neural network that can capture and recall information for future use by the AI system. This is useful in understanding language, for instance.

MACHINE LEARNING

Machine learning (sometimes known as ML) is the method by which machines learn from information to create artificial intelligence systems. 'Deep learning' is the term given to a particularly effective form of machine learning and is now often used to describe complex machine learning AI systems.

MOORE'S LAW

Named after electronics engineer and co-founder of Intel, Gordon Moore, Moore's law describes the ability of the semiconductor industry to rapidly increase the number of transistors that can be squeezed on to a single integrated circuit. Moore predicted in 1965 that the semiconductor industry would be able to double the number of transistors on a microchip each year, for the next ten years. This challenge was subsequently achieved by the semiconductor industry and in 1975 Gordon Moore updated his prediction to say that the semiconductor industry would be able to double the number of transistors every two years. This exponential rate of growth was borne out until relatively recently, but since 2012 it has started to slow down.

MULTI-HEADED SELF-ATTENTION

Multi-headed self-attention is a machine learning technique more commonly called a 'transformer'. It is often used in large language model artificial neural networks as a way to understand the structure of language and to place attention on the key words and phrases in a long string of text.

OVERFITTING

Overfitting is a term that comes from statistics and refers to the fact that if you have only a small number of samples for your statistical analysis, then you may actually be missing key information and therefore your results may be incorrect. In machine learning, if you don't use enough information to train your AI system, it may make mistakes. The AI system might just have 'memorized' the training information rather than being able to 'generalize' from it, and so when it sees new information that is different from the training information it may produce completely inaccurate answers. Getting access to enough training information to train large AI systems is an increasing challenge.

PARAMETER

In an artificial neural network, parameters represent the digest of knowledge that is captured from the information the system learns from. A machine learning method uses each piece of information that the system is learning from to update all the parameters in the artificial neural network. It continues this process with each new piece of training information until the AI system exhibits some intelligent behaviour. As an example, the GPT-4 artificial neural network that is used in the 2023 version of the AI language system ChatGPT contains a total of approximately 1.7 trillion parameters.

QUBIT

The quantum bit, or qubit (pronounced 'cue-bit'), represents the basic unit of quantum information. Unlike a digital bit, which can only represent a '1' or a '0' and is like a switch that is either on or off, a qubit is more like a spinning coin representing both heads and tails (and all possible states in between) at the same time. A qubit represents information by the way that it 'spins', and a quantum computer can perform calculations by getting qubits to 'spin' together.

RECURRENT NEURAL NETWORK (RNN)

A recurrent neural network is a form of artificial neural network used in AI systems that lets the output from certain nodes in the artificial neural network feed back to affect the inputs to the same nodes. This technique creates a simple form of short-term memory. LSTM, or long short-term memory, is a complex form of RNN, and multi-headed self-attention is a further improvement.

RESIDUAL NETWORK

A residual neural network, or residual network (also known as ResNet), is a form of artificial neural network now commonly used as a machine learning method in AI systems. The residual network is modelled on structures found in the brain that allow information to skip a layer so that it can directly influence different layers of parameters in the neural network. When applied to machine learning, this technique has been shown to improve the performance and accuracy of the resulting AI systems. Residual networks were first used to improve convolutional neural networks for image identification tasks but are now also commonly used in other artificial neural networks.

SEMICONDUCTOR

A semiconductor is a special type of material that can either conduct or block an electric current, depending upon its state, allowing it to function like a switch. Silicon is the most common semiconducting element used to build transistor switches in today's integrated circuits (hence

Silicon Valley). The term 'semiconductor' is also now commonly used as shorthand for the integrated circuits found in your laptop.

SHANNON'S LIMIT

Shannon's limit, named after mathematician and communications theorist Claude Shannon, describes the highest theoretical speed at which we can transfer information. This speed limit is 'theoretical' because, in real life, systems can never quite reach this theoretical maximum. The mathematics behind Shannon's limit is today used in all advanced communications systems, such as your Wi-Fi, but can also describe the speed of communication within your brain.

SUPERVISED (AND UNSUPERVISED) LEARNING

Machine learning systems learn from information. In a supervised machine learning system, a label must be added to the information that is used to train the system so that the machine learning system can know what it is meant to find. This method is a bit like flash cards that help children learn: the label DOG is accompanied by a picture of a dog. However, adding labels to the training information is laborious and limits the amount of training information that is available. In unsupervised – or, more correctly, self-supervised – machine learning systems, no labels are required for the training information, making it much more effective.

TRANSFORMER

The term 'transformer' refers to the machine learning technique of 'multi-headed self-attention', which was pioneered by researchers at Google and Toronto University. This technique was later combined with 'generative pre-training' to create the GPT – or 'generative pre-trained transformer' – machine learning method used in systems like ChatGPT.

ACKNOWLEDGEMENTS

The push that finally caused me to sit down and write this book came from a conversation that I had with a member of the British royal family at a business event held at Windsor Castle in October 2021. This discussion was like many others that I'd had, where people were unsure about how AI works, whether it is safe and if it can be trusted. I realized that I must try to share my perspective and some of the knowledge that I have gained from working with leading-edge technology for so many years – and especially what I have learnt about how artificial intelligence thinks.

I am incredibly grateful to all the people in the AI industry who have taken the time to explain different aspects of this amazing technology to me and to share their thoughts on how it will evolve. I apologize for glossing over the underlying details in trying to make an enormously complex subject understandable for anyone who is curious.

I have been extremely fortunate to work with an incredible team at Graphcore, whose knowledge and experience I constantly learn from. Special thanks must go to Simon Knowles, my co-founder and friend, who has been extremely patient in explaining the finer technical points and in sharing a small part of his knowledge. Enormous thanks also go to all the people who took the time to read early drafts of this book and who provided their feedback and encouragement. Some major call-outs are owed to John Kibarian, who read a very early draft and provided enormous encouragement to keep going; to Alex Creswell for his thoughtful feedback; to Paul Neil, who challenged me on an early draft with who the book was written for; to

Paddy for his inputs, and especially to Will for all the time he spent on fixing my spelling (which I still struggle with) and grammar (which remains a mystery) and in providing input on structure. Plus, an enormous thank you to Jeff Ingold, who invested vast amounts of his valuable time helping on early revisions.

I could not have completed this book without the incredible time and effort that have been invested by my agent and editor. They helped me turn what was an extremely rough draft into this final, finished product.

A special mention must go to Helen Wann, who has provided amazing support in my business life for more than a decade and who makes everything possible. And, of course, the biggest thank you goes to Sally for her encouragement, help and constant support. Finally, I must also thank my dog, Red. His doleful, often confused, red setter facial expressions that he shares while on our walks together have been a constant reminder to keep things simple.

PICTURE ACKNOWLEDGEMENTS

INDEX

ABOUT THE AUTHOR

Nigel Toon is the founder of Graphcore and a leading AI entrepreneur. He sits as a non-executive director on the board of the UK
Research and Innovation Council and has sat on the UK's Prime
Minister's Business Council. He has been recognized with numerous
industry awards, being ranked No. 1 on *Business Insider*'s 'UK Tech
100' and named as one of the 'top 100 entrepreneurs in the UK' by
the *Financial Times*. *How AI Thinks* is his first book.